Behavior and Evolution of Birds

Behavior and Evolution of Birds

...

READINGS FROM
SCIENTIFIC AMERICAN
MAGAZINE

Edited by

Douglas W. Mock
University of Oklahoma

W. H. FREEMAN AND COMPANY
New York

Some of the SCIENTIFIC AMERICAN articles in *Behavior and Evolution of Birds* are available as separate Offprints. For a complete list of articles now available as Offprints, write to Product Manager, Marketing Department, W. H. Freeman and Company, 41 Madison Avenue, New York, New York 10010.

Library of Congress Cataloging-in-Publication Data

Behavior and evolution of birds : readings from Scientific American

   magazine / edited by Douglas W. Mock.

      p. cm.

      Includes bibliographical references.

      ISBN 0-7167-2237-2 : $12.95

      1. Birds—Behavior.   2. Birds—
Evolution.   I. Mock, Douglas W.

   II. Scientific American.

   QL698.3.B44   1991

   598.2′51—dc20   91-480
                                             CIP

Printed in the United States of America

1 2 3 4 5 6 7 8 9 0   RRD   9 9 8 7 6 5 4 3 2 1 0

CONTENTS

SECTION III: AVIAN EVOLUTION

Introduction

Note on cross-references to SCIENTIFIC AMERICAN *articles:* Articles included in this book are referred to by chapter number and title; articles not included in this book but available as Offprints are referred to by title, date of publication and Offprint number; articles not in this book and not available as Offprints are referred to by title and date of publication.

Preface

Some people are fascinated by birds from the very start—viscerally stirred by the cacophony of an April dawn's chorus, the languid immobility of a great blue heron peering into a warm summer pond, the travel-poster perfection of goose skeins traversing an October sunset or the chipper mischief of a blue jay industriously relieving a snow-rimmed feeder of its sunflower seeds. Others who find all that pleasant enough but hardly enthralling, may awaken suddenly to the realization that for the birds themselves all these actions are grimly serious business and have been honed relentlessly by natural selection to contribute to eventual reproductive success. The birds, of course, care nothing about our aesthetic sensibilities or our logical insights. They went on with their business long before we tuned in.

Still, we persist. The male songbirds we will hear in spring are the issue of successful singers; preceding males that sang poorly were less successful at acquiring territories and mates, so their traits were passed along less often and eventually vanished. Over the eons of evolutionary time, the same gradual failure must have met any loudly splashing great blue herons, uncoordinated or antisocial geese and lackadaisical jays. This thought provokes us to start asking questions about natural phenomena.

Once we do, our sources of intellectual entertainment are endless and we are hooked.

Whether more people are attracted to birds because of their beauty than for the puzzles they offer is unknown, but it is abundantly clear that some powerful forces draw people to them. Both the aesthetic and intellectual routes sketched in this book rely on the basic conspicuousness of birds; whereas most mammals are too nocturnal or too subterranean for us to encounter them routinely, birds are ubiquitous, readily available for casual contemplation or serious study. Not surprisingly there turn out to be millions of fellow enthusiasts and a standing supply of props for our interest: organizations, books, magazines, scholarly journals, television broadcasts, phonograph records, and tapes.

Capital-"O" Ornithology, *sensu stricto*, may be regarded as encompassing just the professional end of the bird-study spectrum and sometimes seems to have identity problems, perhaps largely because so many other levels of interest in birds exist in our society. To me, Ornithology differs from Bird-Watching by virtue of requiring that its participants be involved in the scientific pursuit of new knowledge (facts) and, better still, new general explanations that can account for entire arrays of facts. Some people are paid (generally unimpressive salaries) for doing this; many others are true amateurs in the noble sense. In addition, of course, there are numerous bird-identification mavens who know a terrifying amount of factual information about the songs, plumages and habits of feathered vertebrates. In my lexicon, then, Ornithology as a label signifies original research and anybody who contributes to our understanding of these animals can claim to be an Ornithologist. Their findings are generally of great interest to fellow bird enthusiasts.

I will now leave the captial-"O", but before abandoning the business of labels, it is worth noting that most modern ornithologists, especially "professional" ones, also have to be some other kind of scientist as well. That is, while the animals under study must have backbones, hot blood and—especially—feathers, the focal questions asked and techniques used derive from one or more concept-based disciplines, such as systematics, physiology or ecology. For many reasons, both good and bad, some of these disciplines are well-represented among ornithologists while others are nearly nonexistent. An example of a bad reason might be the long-held assumption that bird bones, being hollow and relatively fragile, surely disintegrate soon after death, so there is little point in looking for avian fossils. It took a few stubborn paleontologists many years to demonstrate that fallacy. There are a great many bird fossils if one bothers to seek them. Meanwhile, the exciting field of paleo-ornithology lagged in its development until quite recently. A better reason for eschewing birds as research subjects exists in, say, genetics, where such fast-breeding organisms as bacteria and fruit flies divulge their secrets more readily than birds. With good reason, ecology and, increasingly, behavior have attracted bird students because many avian attributes serve particularly well in the pursuit of answers to important questions in these disciplines. (Mayr 1984, Konishi et al., 1989)

Not surprisingly, then, the tenor of ornithology has shifted during the twentieth century from a classical period when the most pressing business was descriptive collecting of specimens and facts to the current scene with its own peculiar and no doubt transient biases. The complexion of tomorrow's ornithology will depend on where the next surges of progress arise. Those surges, in turn, will be dictated by new techniques, new connections between existing leads, new facts that offer unusually rich insights into general patterns and the development of theory.

This book reflects the strong current influence of ecology and behavior. It also shows the infusion of new laboratory techniques from other disciplines (for example, molecular biology and neurobiology) that have been adapted successfully to solving conspicuous and long-standing avian puzzles. This book shows that important bodies of theory—for example, sexual selection, inclusive fitness and optimal foraging—have been tested with birds in the field and lab. Meanwhile, the endless task of describing natural history continues apace.

The chapters have been grouped in three sections: Behavioral Mechanisms (how birds manage to do the amazing things they do); Behavioral Function (why evolutionary forces have shaped their social activities in such ways), and Avian Evolution (the phylogeny and diversity of Class Aves). I readily acknowledge that this particular organization of the chapters reflects my own perspectives as an evolutionary behaviorist who studies feathered vertebrates, but I hope it will be as useful as some other order.

Since the chapters originally appeared as articles in SCIENTIFIC AMERICAN from 1978 to 1990, one of the inherent limitations is that the earliest contributions are no longer as current as one might like. Half of the scientific literature ever published has appeared in roughly the past dozen years and, with well over 100 periodicals devoted mainly or exclusively to the latest finding on birds, this doubling rate may be even faster in ornithology. Therefore, to make these chapters as useful and current as possible, each author has recommended several more recent supplemental readings.

Douglas W. Mock

LITERATURE CITED

Mayr, E. 1984. The contributions of ornithology to biology. BioScience 34:250–255.

Konishi, M., S. T. Emlen, R. E. Ricklefs, and J. C. Wingfield. 1989. Contributions of bird studies to biology. Science 246:465–472.

SECTION

I

BEHAVIORAL MECHANISMS

. . .

Introduction

Upon witnessing an especially impressive feat, be it sleight of hand on a vaudeville stage or a falcon stooping to intercept a panicking pheasant, our first reaction is usually to ask How questions: "How in the world is that done?" This can be a remarkably complicated question, depending on how much detail is desired in the answer. When answering a small child, one might escape with "Well, it takes lots of practice!" But a little reflection makes it clear that a truly complete answer might involve consideration of perceptual systems, motor coordination problems, neurochemistry and so on, in addition to the performer's previous experience at the task.

Because birds do so many extraordinary things, it is not surprising that much research has been devoted to figuring out various answers to "how" questions. The understanding thus gained has made numerous contributions to the study of other animal systems. The five chapters in this section offer a potpourri of behavioral topics. Although they were originally published as fully independent pieces, they mix quite well to provide background on how development influences behavior, with some variation as to emphasis on the physiological bases versus the ecological aspects.

It is likely that no single topic in behavior has received (and, some would say, wasted) more research effort than the so-called nature-nurture problem. Indeed, this is not just a behavior topic; it is a major philosophical current running through human thought and touching on fundamental anxieties about who we are (do we have "free will" or not?) and shows up persistently in our politics and in our literature, from Pygmalion to Tarzan.

In the past the choices were artificially dichotomized so that behavior was ruled to be either instinctive or learned. Bird examples were often given to show this dichotomy. It is quite appropriate then that much of the research in the past quarter-century or so, during which our understanding of the process of behavioral development has deepened well beyond that simple level, has also involved birds. For example one classic "instinct," the habit

of a newly hatched gull chick to peck the red spot on its parent's bill when hungry, has been shown to be far more flexible and modifiable through experience than previously imagined. Overall, the interactions of genetic and nongenetic factors (of which previous experience is but one possible type) are being studied with a growing appreciation of their complexity.

In Chapter 1, "Learning by Instinct," James L. Gould and Peter Marler show that many types of behavior are both innate and learned. They point out that it makes sense for animals to be predisposed to learn certain things more readily than others. In the process they draw many examples from experimental studies of birds (also bees, monkeys and humans), including reinterpretations of behavior patterns that had been classified erroneously according to the old false dichotomy. Their conclusion that different species tend to be smart in areas of great importance to them and remarkably dull in other ways is echoed in several of the following chapters.

In many European songbirds, the timing of long-distance migration is strongly governed by the simple but effective cue of photoperiod at the northern end of the route. In Chapter 2, "Internal Rhythms in Bird Migration" recounted the experiments that revealed a circannual clock set conservatively at about a ten-month cycle, which is recalibrated every summer to produce an effective twelve-month behavioral year. This hormonally mediated system apparently serves to tell the birds when it is time to leave the relatively stable conditions (unvarying day lengths) of equatorial Africa and head north for springtime breeding. Furthermore, different populations within a species are tuned to deal with their own local situation in an appropriate manner and these interpopulational variations apparently have some genetic basis.

An excellent account of how the learning capacities of birds are matched to their ecological requirements is provided in Chapter 3, "Memory in Food-hoarding Birds," by Sara J. Shettleworth. Birds that hoard seeds in numerous small stashes, to be retrieved days or months later, obviously have tremendous need for the ability to remember locations. Some species have remarkable specializations, such as special sublingual pouches that serve as seed suitcases, for this style of life and consequently can take advantage of ephemeral superabundances of food, just as human shoppers stock up on discounted items. Other species that do not live this way have neither the anatomical equipment nor the special mental faculties.

Similarly, there are great between-species differences in the learning of the typical songs and their dialects required for avian social life. Some species get the job done once and for all during their nestling lives, while others remain malleable in this regard, capable of adding new elements every year. It seems logical that the mechanisms underlying how birds acquire song might show similar variability. In Chapter 4, "From Bird Song to Neurogenesis," Fernando Nottebohm reports that recent studies have shown that the prolonged learning capacity is achieved, at least in canaries, by new neural growth and reorganization of the adult brain —a startling discovery. This again underscores the labyrinthine complexity of behavioral development: How exactly would one assign such ability—the performance of novel songs that have been acquired during adulthood on new brain tissues—to one side or the other of the nature-nurture dichotomy?

In Chapter 5, "The Hearing of the Barn Owl," Eric I. Knudsen sets out an age-old natural history problem (how do owls find their prey at night), makes it even tougher (barn owls, which are pretty large, specialize on tiny rodents and so must capture them very efficiently to make up the required numbers) and then dissects the whole story into a manageable set of anatomy, physiology and physics questions. These questions in turn, have been attacked with an elegant series of experiments that show, among other things, that barn owls can rely on acoustic cues alone to guide their strikes in three-dimensional space. In the decade since this review was published, Knudsen's group has carried the barn owl system much farther, showing for example how visual and acoustic inputs are integrated and how vision is used during development to calibrate audition-based localization of prey.

Learning by Instinct

Usually seen as diametric opposites, learning and instinct are patterns — the process of learning, in creatures at all levels of mental complexity, is often initiated and controlled by instinct.

• • •

James L. Gould and Peter Marler
January, 1987

Learning is often thought of as the alternative to instinct, which is the information passed genetically from one generation to the next. Most of us think the ability to learn is the hallmark of intelligence. The difference between learning and instinct is said to distinguish human beings from "lower" animals such as insects. Introspection, that deceptively convincing authority, leads one to conclude that learning, unlike instinct, usually involves conscious decisions concerning when and what to learn.

Work done in the past few decades has shown that such a sharp distinction between instinct and learning — and between the guiding forces underlying human and animal behavior — cannot be made. For example, it has been found that many insects are prodigious learners. Conversely, we now know that the process of learning in higher animals, as well as in insects, is often innately guided, that is, guided by information inherent in the genetic makeup of the animal. In other words, the process of learning itself is often controlled by instinct (see Figure 1.1).

It now seems that many, if not most, animals are "preprogrammed" to learn particular things and to learn them in particular ways. In evolutionary terms

innately guided learning makes sense: very often it is easy to specify in advance the general characteristics of the things an animal should be able to learn, even when the details cannot be specified. For example, bees should be inherently suited to learning the shapes of various flowers, but it would be impossible to equip each bee at birth with a field guide to all the flowers it might visit.

Innately guided learning — learning by instinct — is found at all levels of mental complexity in the animal kingdom. In this chapter our examples will be drawn primarily from the behavior of bees and birds, our respective fields of particular expertise, but the results can be generalized to the primates, even to man. There is strong evidence, for example, that the process of learning human speech is largely guided by innate abilities and tendencies.

TWO THEORETICAL FRAMEWORKS

The distinction often made between learning and instinct is exemplified by two theoretical approaches to the study of behavior; ethology and behaviorist psychology. Ethology is usually thought of as the study of instinct. In the ethological world view most animal behavior is governed by four

Figure 1.1 INSTINCTIVELY GUIDED LEARNING enables the cuckoo (*left*) to parasitize other species of birds (in this case a hedge sparrow). Cuckoos lay their eggs in the nests of other birds. When the cuckoo hatches, its begging tricks those parents into accepting it as their own young. While it is still in the nest, the cuckoo (if it is female) must somehow learn to recognize the species of its foster parents so that it can lay its own eggs in the nests of that species. In this learning process the cuckoo is guided by instinct to ignore a world full of distracting information in order to focus on the details that must be memorized.

basic factors: sign stimuli (instinctively recognized cues), motor programs (innate responses to cues), drive (controlling motivational impulses) and imprinting (a restricted and seemingly aberrant form of learning).

Three of these factors are found in the egg-rolling response of geese, a behavior studied by Konrad Z. Lorenz and Nikolaas Tinbergen, who together with Karl von Frisch were the founders of ethology. Geese incubate their eggs in mound-shaped nests built on the ground, and it sometimes happens that the incubating goose inadvertently knocks an egg out of the nest. Such an event leads to a remarkable behavior. After settling down again on its nest, the goose eventually notices the errant egg. The animal then extends its neck to fix its eyes on the egg, rises and rolls the egg back into the nest gently with its bill. At first glance this might seem to be a thoughtful solution to a problem. As it happens, however, the behavior is highly stereotyped and innate. Any convex object, regardless of color and almost regardless of size, triggers the response; beer bottles are particularly effective.

In this example the convex features that trigger the behavior are the ethologists' sign stimuli. The egg-rolling response itself is the motor program. The entire behavior is controlled by a drive that appears about two weeks before the geese lay eggs and persists until about two weeks after the eggs hatch. Geese also exhibit imprinting: during a sensitive period soon after hatching, goslings will follow almost any receding object that emits an innately recognized "kum-kum" call and thereafter treat the object as a parent.

Classical behaviorist psychologists see the world quite differently from ethologists. Behaviorists are primarily interested in the study of learning under strictly controlled conditions and have traditionally treated instinct as irrelevant to learning. Behaviorists believe nearly all the responses of higher animals can be divided into two kinds of learning called classical conditioning and operant conditioning.

Classical conditioning was discovered in dogs by the Russian physiologist Ivan P. Pavlov. In his classic experiment he showed that if a bell is rung consistently just before food is offered to a dog, eventually the dog will learn to salivate at the sound of the bell. The important factors in classical conditioning are the unconditioned stimulus (the innately recognized cue, equivalent to the ethological sign stimulus, which in this case is food), the unconditioned response (the innately triggered behavioral act, equivalent to the ethological motor program, which in this case is salivation) and the conditioned stimulus (the stimulus the animal is conditioned to respond to, which in this case is the bell). Early behaviorists believed any stimulus an animal was physically capable of sensing could be linked, as a conditioned stimulus, to any unconditioned response.

In operant conditioning, the other major category of learning recognized by most behaviorists, animals learn a behavior pattern as the result of trial-and-error experimentation they undertake in order to obtain a reward or avoid a punishment. In the classic example a rat is trained to press a lever to obtain food. The experimenter shapes the behavior by rewarding the rat at first for even partial performance of the desired response. For example, at the outset the rat might be rewarded simply for facing the end of the cage in which the lever sits.

Later the experimenter requires increasingly precise behavior, until the response is perfected. Early behaviorists thought any behavior an animal was physically capable of performing could be taught, by means of operant conditioning, as a response to any cue or situation.

CHALLENGES TO BEHAVIORISM

By 1970 several disturbing challenges to the behavioristic world view had appeared. The idea that any perceptible cue could be taught, by classical conditioning, as a conditioned stimulus was dealt a severe blow by John Garcia, now at the University of California at Los Angeles. He showed that rats could not associate visual and auditory cues with food that made them ill, even though they could associate olfactory cues with such food. On the other hand, he found that quail could associate not auditory or olfactory cues but visual ones — colors — with dangerous foods. Later work by other investigators extended these results, showing, for example, that pigeons readily learn to associate sounds but not colors with danger and colors but not sounds with food. The obvious conclusion was that these animals are predisposed to make certain associations more easily in some situations than in others.

The same kind of pattern was discovered in experiments in operant conditioning. Rats readily learn to press a bar for food, but they cannot learn to press a bar in order to avoid an electric shock. Conversely, they can learn to jump in order to avoid a shock but not in order to obtain food. Similarly, pigeons easily learn to peck at a spot for a food reward but have great difficulty learning to hop on a treadle for food; they learn to avoid shock by hopping on a treadle but not by pecking. Once again it seems that in certain behavioral situations animals are innately prepared to learn some things more readily than others.

The associations that are most easily learned have an adaptive logic. In the natural world odor is a more reliable indicator than color for rats (which are notoriously nocturnal) trying to identify dangerous food; the color of a seed is a more useful thing for a pigeon to remember than any sounds the seed makes. Similarly, a pigeon is more likely to learn how to eat novel seeds if it experiments on food with its beak rather than with its feet. Animals that have innate biases concerning which cues they rely on and which procedures they attempt are more likely to ignore spurious cues, and they will learn faster than animals without inherent biases. The idea that animals are innately programmed to attend to specific cues in specific behavioral contexts and to experiment in particular ways in other contexts suggests a mutually reinforcing relation between learning and instinct. This relation helps to explain the once anomalous phenomenon of imprinting and to reconcile the approaches of behaviorists and ethologists.

INSTINCTIVE LEARNING IN BEES

The convergence of the two perspectives is illustrated by the ways honeybees learn about flowers. Bees inherently learn certain characteristics of a flower more easily than they learn others. Perhaps even more significant, once bees have acquired knowledge about a flower the ways in which they organize and refer to that knowledge are entirely instinctive.

Bees make their living by collecting nectar and pollen. Both of these essential foods are found in flowers, which offer them as a bribe to attract pollinating insects. Bees recognize flowerlike objects instinctively: they land spontaneously on small, brightly colored objects that have a high spatial frequency, or ratio of edges to unbroken areas (the spatial frequency of an object is high, for example, if the object has petals), and centers that (like the center of a flower) absorb ultraviolet light and so appear dark to bees.

Although bees recognize flowerlike objects innately, they have to learn which of those objects are likely to hold food (see Figures 1.2 and 1.3). The initial flowerlike characteristics constitute an unconditioned stimulus: a set of sign stimuli. They trigger the unconditioned responses of landing and probing with the proboscis, behaviors that represent two innate motor programs. If a flowerlike object rewards a bee with food, the flower's specific characteristics may be learned—imprinted—as conditioned stimuli.

The first thing honeybees learn about a flower is its odor. Von Frisch showed early in his career that after a bee has been trained by being fed at a feeder that has a particular odor, it selects flowers of like odor from among hundreds of alternatives. Randolf Menzel of the Free University of Berlin showed that even one training visit is enough to teach the bee to choose the same odor 90 percent of the time in later visits; after only three training visits the rate of success is higher than 98 percent [see "Learning and Memory in Bees," by Randolf Menzel and Jochen Erber; SCIENTIFIC AMERICAN, July, 1978; Offprint 1395]. Bees do not learn all odors with equal ease. Nonfloral odors take longer to learn, although it is unclear whether the bias results from insensitivity to inappropriate odors or from some problem in remembering them.

The next thing honeybees learn about a flower is its color. Menzel has shown that roughly three training visits to flowers of the same color are necessary before bees select that color over an alternative color 90 percent of the time. After about 10 training trips the bees choose the correct color more than 95 percent of the time. As with odors, bees do not learn all colors equally quickly, but enough is known about the vision of bees to rule out the possibility that this bias is based on an unevenness in the bees' ability to see different colors.

Honeybees also learn the shapes and color patterns of flowers, but they need more training visits in order to reach the level of 90 percent accuracy in remembering shape; about five or six visits suffice to enable them to distinguish a square "flower" (a plastic square containing a feeder) from a triangular one.

As with odor and color, bees inherently prefer some shapes over others. They particularly prefer busy patterns to simple ones. Until recently most investigators had thought bees do not remember a pattern as a picture (unlike human beings, and vertebrates in general) but rather as a list of defining characteristics, in much the same way as advertisements for real estate often depend not on photographs but on verbal lists: "Red Cape Cod, three bedrooms, two baths, detached garage." Such a list might enable bees to distinguish among species of flowers, and it would not require as large and complex a central nervous system as a picture memory would. Recent experiments by one of us (Gould) indicate, however, that bees do store low-resolution pictures of flowers.

Bees learn many things about flowers, but there are some cues that cannot be stored as part of flower memory even though bees can learn them in other behavioral contexts. For example, honeybees are famous for their exquisite sensitivity to polarized light (by which they navigate), but they cannot learn the polarization patterns of flowers. They are also adept at learning which way a hive faces (to the point where rotating the hive by 90 degrees leaves most forages unable to find the entrance until other bees provide strong chemical cues), but they will

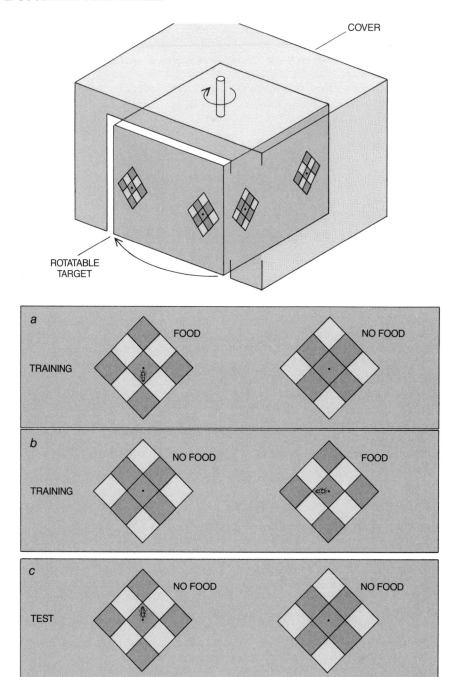

Figure 1.2 TRAINING AND TESTING APPARATUS teaches bees to land on particular targets and checks their ability to remember the targets. Pairs of targets are arranged on each side of a rotatable box (*top*). To train the bee, the investigator supplies food to one target and lets the bee feed there (*a*). To keep the bee from simply memorizing the location of the target bearing food, the box is rotated so that the bee can also be trained on a second pair of targets that is a mirror image of the first (*b*). After the bee has been fed about 10 times, the box is rotated to expose targets, both of which contain no food (*c*), and the bee is watched to see which it chooses to land on first.

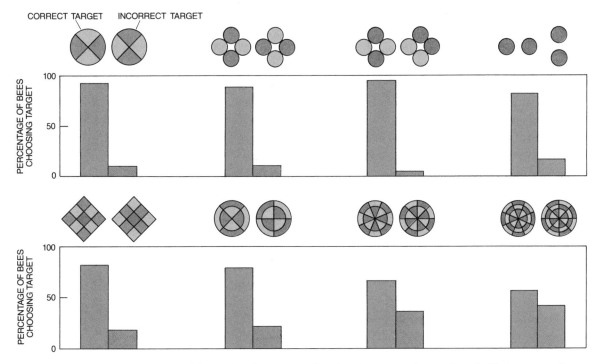

Figure 1.3 TARGETS provided evidence that bees remember a flower as a picture, not as a list of characteristics. The targets in each pair differ from each other only in the arrangement of their components (often one is merely a rotation of the other), and so they would be represented by identical lists of characteristics. Bees would be able to distinguish the targets in a pair only if they remember a target as a picture. Bees remembered the target they had been trained on except when it was very intricate; their memory may not have fine enough resolution to distinguish such targets. The targets shown are a few of many used in experiments.

not learn in what direction a freestanding flower faces.

ORGANIZATION OF BEE KNOWLEDGE

The cues bees do remember about flowers, such as odor, color and pattern, are not remembered with equal weight. For example, if a bee that has been trained to feed at a peppermint-scented blue triangular target is presented with a choice between an orange-scented blue triangular target and a peppermint-scented yellow circular target, it will inevitably choose the peppermint-scented target even though that target has neither the color nor the shape the bee has been trained on. It is only when two targets have the same odor that bees pay much attention to color or shape; under those conditions color takes precedence over shape. This hierarchy corresponds to the relative reliability of the cues in nature (see Figure 1.4). The odor of a flower is usually constant,

whereas color can fade or appear different under different lighting conditions, and shape changes with damage from wind and herbivores, and even with viewing angle.

The hierarchy is an important factor in the organization of the bees' memory, but there is an even more important organizational element: the time of day at which each flower provides nectar. Bees learn the time at which food is available from each flower more slowly than they learn odor, color or shape, but once they have learned it, that knowledge serves to organize their use of the rest of their memory.

The organizational role of time was clearly shown by Franz Josef Bogdany of the University of Würzburg. For several days he trained a set of foragers to feed at two different feeders at differing times of day. From 10:00 to 11:00 A.M., for example, he fed them at a peppermint-scented blue triangular feeder; from 11:00 A.M. until noon he fed them at an

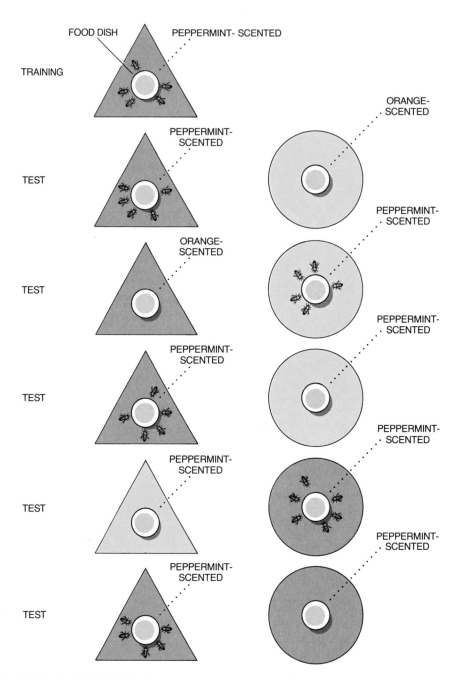

Figure 1.4 HIERARCHICAL STRUCTURE of bees' memory is revealed in a series of tests. Bees were trained to land on peppermint-scented blue triangles. Their training was confirmed by a test in which they preferred a peppermint-scented blue triangle to an orange-scented yellow circle, even when both targets bore full food dishes. Then they were offered an orange-scented blue triangle and a pep-permint-scented yellow circle. They chose the latter, showing they are more likely to be guided by the memory of a scent than by the memory of a shape or color. Later tests, interspersed with controls, showed bees are more likely to rely on a remembered color than a remembered shape.

orange-scented yellow circular feeder that was placed on the site the blue feeder had occupied. One day he put both feeders out at 9:00 A.M. and noticed an interesting pattern. Trained foragers began to appear at the blue feeder at about 9:45. They foraged exclusively at that feeder for about an hour. At roughly 10:45 some foragers began to shift to the yellow feeder, and by 11:15 the blue feeder—which was still full of food—was completely abandoned.

Bees behave as though they have an important book by which they schedule their visits; no more than one entry can be made for any specific time. The resolution of the book is about 20 minutes; that is, bees cannot remember two separate appointments if they are less than 20 minutes apart. Bees have been able to remember as many appointments as experimenters have tried to teach them. The standing record, set by R. Koltermann of the University of Frankfurt, is nine appointments in eight hours.

Another experiment by Bogdany shows finer details in the structure of honeybee memory. After days of being trained to the peppermint-scented blue triangular feeder, bees were presented with an orange-scented blue triangular feeder. The foragers learned the new odor in one visit, but they completely forgot the color and shape, even though these characteristics had not been changed. On the other hand, when bees were trained to an odorless blue triangle and then presented with a peppermint-scented blue triangle, they learned the new odor without forgetting the color and shape. Apparently the appointment book has an entry for each cue; the entries are structured in such a way that blanks can always be filled in but that if even one item is changed, the entire entry is erased.

Such results suggest that honeybees, guided to particular targets by instinctively recognized cues, memorize certain specific features of the targets and store that memory in a "prewired," hierarchical memory array. The cues that are memorized, the speed with which each cue is memorized and the way the memorized data are stored are all innate characteristics of the bee.

LEARNING ABOUT ENEMIES

Animals need to learn many things other than how to find food. For example, they must learn how to recognize and respond to various kinds of predators and enemies. For some animals it is enough merely to identify a very general class of predators. Flying moths and crickets automatically begin evasive maneuvers when they hear the high-pitched sounds characteristic of hunting bats. Other animals must be able to make finer distinctions among friends and potential foes. Nesting birds are a particularly apt example. They must learn to distinguish harmless birds, such as robins, from birds such as crows and jays, which hunt for eggs and nestlings. The "fill in the blanks" strategy adopted by bees as they learn about flowers is also applicable to this kind of learning.

When nesting birds detect nest predators, they attack en masse, a phenomenon commonly known as "mobbing." How do birds know whom to mob and whom to ignore? Eberhard Curio of the University of Ruhr has shown that the process of learning which species to mob is innately guided.

In Curio's experiments groups of birds (most often European blackbirds) were kept in separate cages. Between the cages was placed a rotatable box with four compartments. At any given time the birds in one cage could see into only one compartment of the rotatable box, while the birds in the other cage saw a different compartment (see Figure 1.5). The birds could see into each other's cages.

Curio began by rotating the central box to present a stuffed Australian honeycreeper—a harmless species—to each cage. The live birds showed no reaction. He then put a stuffed owl in one compartment and a honeycreeper in the opposite compartment. When the box was rotated so that each model was in view of one of the two sets of birds, the birds in the cage exposed to the owl began to emit the species' innate mobbing call and tried to attack the model. The other group observed the mobbing for a moment and then, responding to this powerful set of sign stimuli, began trying to attack the stuffed honeycreeper, at the same time emitting the mobbing call. On later occasions this group of birds always tried to mob honeycreepers, a species they had never seen attack a nest. Curio found that the baseless aversion to honeycreepers was passed on from generation to generation. The young birds learned to mob honeycreepers by watching their parents. In later experiments Curio was able to teach his birds to mob bottles of laundry detergent.

There is good reason to think variations of this strategy of learning about enemies are at work in many mammalian species as well as in birds. Perhaps the most elaborate version is found in vervet monkeys. As was shown by Robert M. Seyfarth,

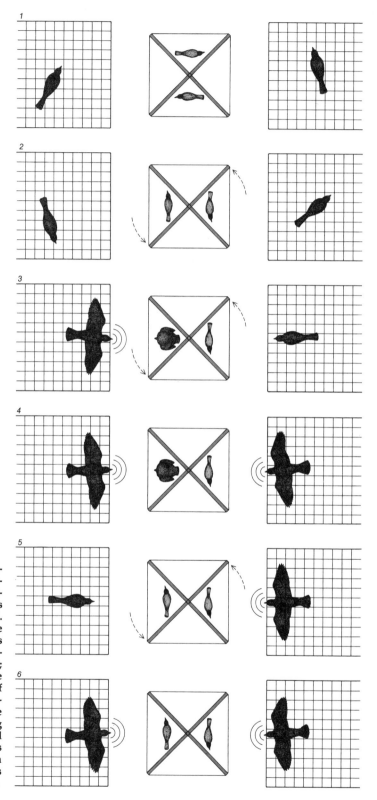

Figure 1.5 MOBBING BEHAVIOR of the European blackbird demonstrates the instinctive way the species learns to recognize predators. Between the cages in which the birds sit is a rotatable, four-chambered box (1). Each bird can see only one chamber of the box, but it can also see into the other bird's cage. First each bird is shown a stuffed Australian honeycreeper, a harmless species (2); neither bird shows any interest. Then one bird is shown a stuffed owl (a predator of small birds) and the other is shown a honeycreeper. The bird shown an owl tries to chase it away and gives the characteristic "mobbing call" (3). The other bird at first watches and then (4) joins in the mobbing behavior. It has learned to mob honeycreepers. When both birds are shown honeycreepers (5), it teaches the other to mob honeycreepers as well (6).

Dorothy L. Cheney and one of us (Marler), vervets have special alarm calls for each of four kinds of predators: aerial predators such as eagles, four-legged predators such as leopards, predatory primates such as baboons, and snakes. Each alarm call elicits a different kind of response. For example, an eagle alarm sends vervets on the ground toward cover and causes those in the exposed tops of trees to drop like stones into the prospective interior, whereas a snake call is ignored by vervets in trees but induces those on the ground to rear up on their hind legs and scan the ground around them.

Young vervets instinctively emit alarm calls in response to a wide but specific range of stimuli. For example, any object within certain size limits moving in free space at a certain angular velocity causes the young monkeys to give the eagle call; the call can be elicited by a stork or even a falling leaf. With time the infants learn which species cause the adults to call. Hence vervets growing up in one region might learn to give alarm calls on seeing baboons, leopards and a certain species of eagle, whereas those in another region might react to human beings, hunting dogs and a certain species of hawk. Like the bees' system of learning about flowers, this innate system is efficient for learning essential information about predictably unpredictable situations: predictable kinds of threat posed by animals whose exact species cannot be predicted.

SONG LEARNING IN BIRDS

Another task an animal must perform that often requires learning is recognizing others of its own species. Perhaps the richest and best understood use of learning in species recognition is the learning of songs by birds. All birds have a repertoire of perhaps one or two dozen calls that are innately produced and recognized. These calls need not be learned and can be produced even by birds hatched and reared in isolation. Several kinds of birds also have more complex vocal patterns—songs for attracting mates and defending territory—that must to some extent be learned from adults of the same species.

The white-crowned sparrow, which has been studied extensively by one of us (Marler), is a good example (see Figure 1.6). Adults of this species produce a three-part of four-part courtship song rich in melodic detail. Different individuals produce recognizably different songs, but the organization of the song is common to the species. The song produced by each male white-crown is similar to (but not identical with) the songs heard near the place it was reared. (There are actually local dialects.)

Experiments in manipulating the sensory experience of young sparrows have revealed much about the organization of the process of song learning. A bird kept in auditory isolation, for example, begins to produce and experiment with song notes by the time it is about a month old. This period of experimentation, known as subsong, waxes and wanes for roughly two months. By about the bird's 100th day it "crystallizes" its song into a form that will not change significantly; the song is highly schematic, but it bears many of the basic features of normal adult white-crown song. Such experiments show that the chick is born with a basic innate song, which it learns to elaborate when it is raised in the wild.

In another experiment we play tape-recorded songs of other species to isolated young birds. Such songs have little effect on the final, crystallized song produced by the bird (although Luis Baptista of the California Academy of Sciences has shown that a live tutor can sometimes successfully indoctrinate young white-crowns). On the other hand, when we play a medley of tape recordings, one of which reproduces a real white-crown song, the young male somehow manages to pick out the white-crown song and learns to produce a tolerable imitation. If it is to produce a perfect imitation, the bird must hear the song before it is about seven weeks old. (The actual period varies with experimental conditions.) The "window" for learning (the time in which the drive to learn is high) is called the sensitive period.

Taken together, these experiments show that learning of the local song dialect in the white-crowned sparrow is innately controlled, irreversible and restricted to a sensitive period; these are exactly the characteristics of classical imprinting. Starting with some form of innate song, the young male depends on innately recognized cues to trigger the learning process and later learns to imitate the more elaborate, memorized song.

The actual process of developing the bird's own version of the local dialect seems to be done by trial-and-error learning. In the case of the white-crowned sparrow, weeks or months may pass between the ends of the sensitive period (during which the bird memorizes the song) and the birds first experiments with recognizable imitations, which come at the end of subsong. Masakazu Koni-

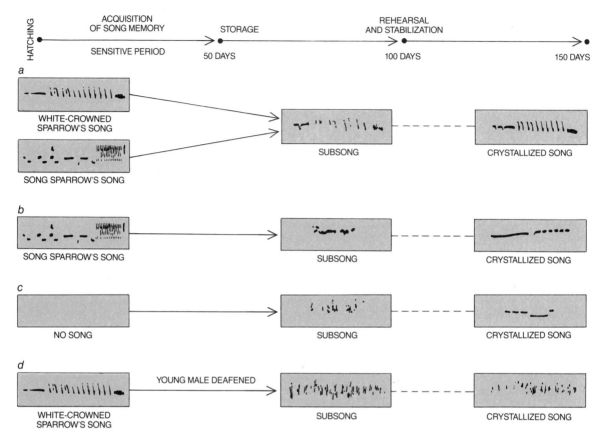

Figure 1.6 SONG LEARNING in the white-crowned sparrow exhibits great specificity: young males preferentially learn their own species' song. If a young male white-crown is played tape recordings of adult white-crown song and adult song-sparrow song (*a*), it begins a period of experimentation known as subsong and then produces a crystallized song very similar to the white-crown song it has heard. If it is played only a tape recording of song-sparrow (*b*) it will not learn the song: it still goes through subsong, but its final, crystallized song does not resemble either the song-sparrow song or the white-crown song. A bird that is played no song (*c*) also learns nothing. If the young bird hears a white-crown song but is deafened before subsong begins (*d*), it is unable to learn how to produce the song it heard.

shi has demonstrated that when a young male white-crown that has been exposed to white-crown song is deafened before any crystallization, it never sings anything melodic—not even its innate song. Apparently a bird must experiment with its beak, syrinx and pulmonary muscles, listening to the sounds that result from various manipulations and trying to match them to its mental record. During the progression from subsong to crystallization the bird shapes its song to match the record stored in its brain, whether that record consists of only the rough, innate song or of a song memorized during the sensitive period. Konishi found that by the time the bird has crystallized its song the singing has become so routine that deafening has little or no effect.

THE CUES WITHIN ADULT BIRDSONGS

How do the young songbirds pick out their own species' song from a world full of sounds? What specific cues do the sparrows rely on to determine which song to learn? One of us (Marler) and his colleagues have investigated this question in experiments with swamp sparrows and song sparrows, two species that nest within earshot of one another (see Figure 1.7).

Of the two species, swamp sparrows have the simpler song. It consists of a single series of regularly repeated syllables; the kinds of syllables vary from theme to theme, bird to bird and region to region. The song sparrow's song is more complex: it consists of at least four types of syllables, often beginning with an accelerating trill. Although the innate songs of the two species reflect some of these structural differences, the syllables from which the innate songs are constructed are much simpler.

The auditory cues that guide learning might include elements of the syllables themselves, elements of the tempo and phrase structure or elements of both. As a first stage in determining which of these elements are important, we gave hand-reared sparrows of both species a chance to learn from tapes of their own species or tapes of the other species. As we had expected, the birds learned almost exclusively from tapes of their own species. The rare cross-species imitations are important, however: they show that songs of other species can physically be sung, and that the normal tendency not to learn the song of another species comes from the birds' inattentiveness to such songs rather than from inability to produce them.

To investigate further the role of different aspects of song structure in learning preferences, we put together synthetic training songs that varied in structure and played them to young birds of both species. For example, we fabricated one song from a slow, steady repetition of song-sparrow syllables and another from a slow, steady repetition of swamp-sparrow syllables. The tempo of these songs was like that of swamp-sparrow song. Swamp sparrows readily learned the steady song with swamp-sparrow syllables but not the other song, and song sparrows learned only the song containing song-sparrow syllables. these results indicate that cues lie within syllables.

By itself that experiment did not indicate whether or not tempo and phrase structure might be important as well. To investigate this issue we synthesized a variety of other songs. The new songs were created from syllables of both species, but the syllables were organized in a variety of patterns. Some of the patterns were like those of the swamp sparrow and others were like those of the song sparrow. One of the songs, for example, was made up of swamp-sparrow syllables but had the accelerating rate of delivery characteristic of song sparrows. Would young swamp sparrows reject the song because it had the wrong tempo? Or would they accept it because it had the correct sign stimulus for learning (the right syllables) and so sing a abnormally paced song? In fact, the young swamp sparrows did neither. They learned the syllables of the song, but in their own singing they actually changed the tempo, so that they delivered the learned syllables at the constant rate typical of their own species.

In another variation we fabricated songs even more similar in structure to song-sparrow songs. Each song had two segments; each segment consisted of a different type of syllable, and in one segment the syllables were delivered at an accelerating rate. When the songs were made up of different kinds of swamp-sparrow syllables, young swamp sparrows learned to sing a steady repetition of one of the two syllable types, regardless of the temporal pattern in which the syllables had been presented. Swamp sparrows thus seem to focus entirely on the syllabic structure when searching for cues, paying scant attention to the organization of the song as a whole.

Song sparrows are different. They are readier to accept the alien syllables of the swamp sparrow if the syllables are presented in song models that have complex phrase structures (although they reject swamp-sparrow syllables when they are delivered as a steady tempo). The two attributes—syllable type and syntactic structure—apparently have additive effects.

These experiments show that although the song sparrow and the swamp sparrow are closely related, the innate mechanisms that control learning in the two species are different. No doubt the white-crowned sparrow is different in its own way.

SPEECH LEARNING IN HUMANS

The learning of songs in birds has a number of parallels with the learning of speech in human beings. In swamp sparrows, song learning involves the innate recognition of certain elements in the species-specific syllables. There is now abundant evidence that human infants innately recognize most or all of the more than two dozen consonant sounds characteristic of human speech, including consonants not present in the language they normally hear. The innate ability to identify sign stimuli present in consonants confers several advan-

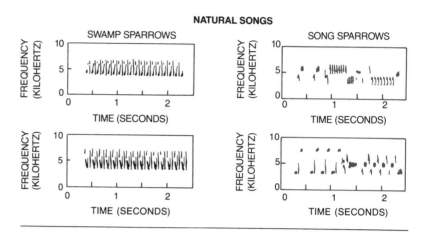

NATURAL SONGS

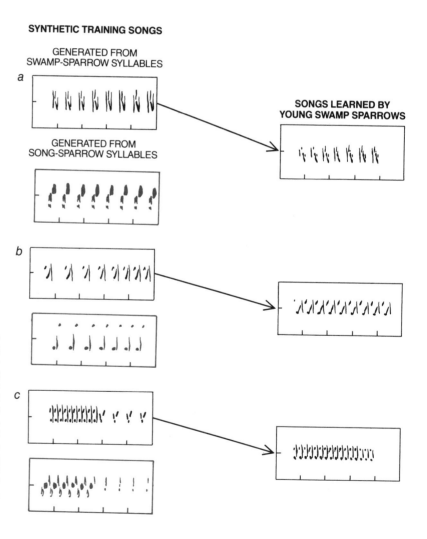

SYNTHETIC TRAINING SONGS

GENERATED FROM
SWAMP-SPARROW SYLLABLES

a

SONGS LEARNED BY
YOUNG SWAMP SPARROWS

GENERATED FROM
SONG-SPARROW SYLLABLES

b

c

Figure 1.7 SYNTHETIC SONGS reveal that young swamp sparrows rely on cues within individual syllables when determining what song to learn. Natural swamp-sparrow song (*top left: two shown to display variation from bird to bird*) consists of steady repetition of one syllable, while song sparrow song (*top right*) is polysyllabic and variable in rate. Swamp sparrows prefer their own species' notes, regardless of whether the rate was steady (*a*), accelerating (*b*), or given the temporal pattern of song sparrows (*c*).

tages: it allows the infant to ignore a world full of irrelevant auditory stimuli in order to focus on speech sounds, it starts the child on the right track in learning to decode the many layers of meaning buried in the immensely complex and variable sounds of speech, and it provides an internal standard for the child to use in judging and shaping speech sounds.

Another aspect of human speech learning parallels the subsong phrase, the period during which birds of species that learn to sing begin to experiment with sound production. The subsong phase begins right on schedule even if the bird has been deafened (although such birds learn nothing from their vocal experimentation). Human infants also have a phase of babbling, in which they develop, through trial-and-error learning, the ability to produce the set of consonants found in their own language. As with birds, babbling begins and ends on schedule even in deaf children.

Birds have an innate, templatelike structure that specifies the rules for producing syllables in song. On a vastly different scale, there is good reason to believe that the rhythms in which words and sentences are assembled in speech and the set of rules known as grammar (in particular the division of words into such categories as nouns, verbs, adjectives and adverbs) are at some deep level also innate. This idea, argued most persuasively by Noam Chomsky of the Massachusetts Institute of Technology, helps to explain why the learning of speech proceeds so easily compared with the learning of such inherently simpler tasks as addition and subtraction.

SOPHISTICATED LEARNING

Although much of animal learning (and probably more of human learning than is yet suspected) is innately guided by learning programs, much human behavior clearly cannot be explained so simply. For example, imagining a solution before exploring it physically is a behavior outside the two traditional forms of learning originally studied by behaviorists. This kind of cognitive learning, called cognitive trial and error, comes much closer than programmed learning to one's intuitive sense of what intelligence is. It requires the ability to recall and combine separate bits of learned information and from such mental recombination to formulate new behavioral solutions.

The first evidence that animals have such an ability came in 1948 in a series of experiments by Edward C. Tolman of the University of California at Berkeley. In one experiment Tolman allowed rats to explore a maze and two had alternative goals, a white box and a black box, both containing food; rats learned the routes to both boxes, and chose them with equal frequency. Later Tolman took the rats to another room, where a black box and a white box were put side by side, and he gave them a shock when they entered the black box. When the rats were released in the maze the next day, they entered only the white box. Tolman concluded that they had combined information from the two entirely different experiments, generalizing about black boxes and remembering that one route led to a black box. Tolman also found that rats have an ability to form mental maps of familiar areas and from the maps to plan novel routes. Tolman's finding was subsequently confirmed and explored by David S. Olton of Johns Hopkins University [see "Spatial Memory," by David S. Olton; SCIENTIFIC AMERICAN, June, 1977].

The ability to form maps is by no means limited to rats and human beings. Emil W. Menzel of the State University of New York at Stony Brook has found the same ability in captive chimpanzees; John R. Krebs of the University of Oxford and Sara J. Shettleworth of the University of Toronto have shown that seed-caching birds can form cognitive maps registering the locations of hundreds of hidden seeds (see Chapter 3, "Memory in Food-hoarding Birds," by Sara J. Shettleworth).

MENTAL MAPS AND CATEGORIES

In an effort to determine how common this sophisticated cognitive ability is, one of us (Gould) investigated whether honeybees also have mental maps (see Figure 1.8). When bees travel a familiar route, they rely on prominent landmarks. The usual explanation of how bees use landmarks is that they remember the series of landmarks encountered en route to each site and can refer to the landmarks only in the same way as Hansel and Gretel would have referred to their trail of breadcrumbs. In that case bees would have no idea how one set of landmarks leading to one site is related spatially to the set that leads to any other site.

We tested whether bees do navigate this way or whether they in fact put landmarks in the context of a mental map of their home area. We trained certain bees to feed in one area and then, on subsequent days, captured them as they flew from the hive to

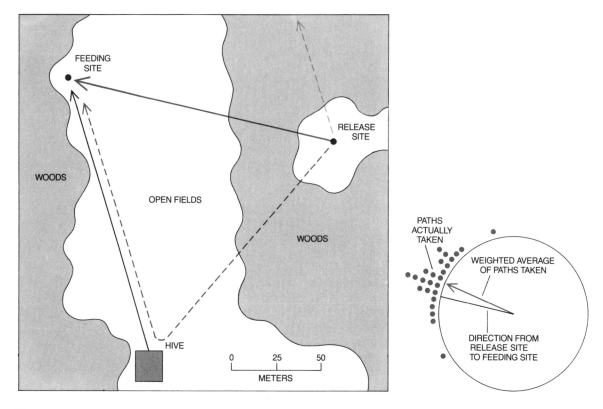

Figure 1.8 ABILITY OF BEES to construct mental maps was demonstrated in experiments in which bees were trained to feed at a specific site (*top*). After the training period a number of bees were caught as they left the hive to go to the feeding site. They were transported, in the dark, to another site (*right*) and released. If the bees had not noted their new surroundings, they might have flown off at the compass bearing that would normally have taken them from the hive to the feeding site (*light colored broken arrow*). If bees remember routes simply by remembering strings of landmarks, they might have followed landmarks back to the hive and then to the feeding site (*dark colored broken arrow*). Instead the bees used their knowledge of the area to devise a new route from the release site directly to the feeding site (*solid colored arrow*).

that area and carried them (in the dark) to another location.

We thought that when the bees were released in the new area, they might adopt any of several courses of action. They could be disoriented and fly at random. Alternatively, they could fail to understand that they had been displaced; they would then fly off at the same compass bearing they would normally follow to get to the food site from the hive. If bees can navigate only by specific strings of landmarks, they might recognize the landmarks around the new location as part of a route different from the one they were on when they were captured; they would then follow that route back to the hive and from there fly to the food site. Finally, if bees

do have cognitive maps, they should be able to determine where they were in relation to the food site and to select the appropriate bearing to reach the food, even though they had never flown from the hive to the food site before by such a roundabout route.

We found that bees overwhelmingly followed the last of these alternatives: when the area to which they were displaced was within their home area (the four or so square kilometers immediately surrounding the hive), they flew directly to the food site. It seems, then, that for bees cognitive map making is an innate part of route learning.

Another sophisticated ability involved in the process of learning is the formation of abstract concepts and categories. Is this ability found in animals? One suggestive hint comes from the work of Richard J. Herrnstein of Harvard University. He showed thousands of slides to laboratory-reared pigeons and rewarded them when they pecked at slides in which some specific kind of object, say a tree, was in the picture; these birds, of course, had never seen a real tree. The birds learned the task remarkably quickly, which suggests that they had a strong innate disposition to form generalized conceptual categories. When they were later tested with slides showing new species of trees, the birds reliably picked out the slides with trees, including some slides the experimenters had at first thought were treeless. The birds' occasional errors were also revealing: they sometimes identified telephone poles and television antennas as trees.

Students of human language acquisition have long known that children automatically form conceptual categories for the new words they learn. Chairs, tables and lamps are organized into a "furniture" category and the category of "chairs" is subdivided into subordinate categories such as rocking chairs and armchairs. Such categorization is essential to the rapid acquisition of words, and word storage in the brain is probably organized as a categorized filing system. The effects of small strokes, which can kill small regions of the brain, seem to reflect such a system: their victims sometimes lose an entire category of words, for instance the names of flowers.

It seems reasonable to propose that the drive to categorize is innate in at least some species. Perhaps it is the ability to make and manipulate categories that underlies the ability of animals to perform cognitive trial and error: to evaluate alternatives and formulate simple plans.

A NEW SYNTHESIS

The emerging picture of learning in animals represents a fundamental shift from the early days of behaviorism, when animals were supposed to be limited to learning by classical conditioning and operant conditioning and were expected to be able to learn any association or behavior by those processes. It is now understood that much learning, even though it is based on conditioning, is specialized for the learning of tasks the animal is likely to encounter. The animal is innately equipped to recognize when it should learn, what cues it should attend to, how to store the new information and how to refer to it in the future. Even the ability to categorize and perform cognitive trial and error, a process that may be available to the higher invertebrates, may depend on innate guidance and specialization—specialization that enables the chickadee, with its tiny brain, to remember the locations of hundreds of hidden seeds, whereas human beings begin to forget after hiding about a dozen.

This perspective allows one to see that various animals are smart in the ways natural selection has favored and stupid where their life-style does not require a customized learning program. The human species is similarly smart in its own adaptive ways and almost embarrassingly stupid in others. The idea that human learning evolved form a few processes, which are well illustrated in other animals, to fit species-specific human needs helps to bring a new unity to the study of animal behavior and a new promise for understanding human origins.

Internal Rhythms in Bird Migration

Migratory birds have a clock that tells them when to begin and end their flight. It is based on rhythms with a period of about a year. Remarkably, the clock also helps the birds to find their destinations.

• • •

Eberhard Gwinner
April, 1986

The sight of a flock of birds migrating south in the fall or north in the spring hardly ever fails to evoke a sense of wonder. The flight may be the orderly aerobatics of a V of Canada geese or the ragtag progress of a group of starlings. Whatever its details, the overwhelming impression conveyed to the observer is that of a powerful inner impulse. The birds do not hesitate in their flight; they travel smoothly and unerringly toward a goal far out of the viewer's sight. Where does the impulse come from that guides the birds toward warmer climates in winter and brings them back to their northern breeding grounds in the spring?

There are two possible kinds of answer to the question. The first is that the impulse springs from factors originating outside the bird, in its environment. Among those that come to mind immediately are the changes in temperature and in the duration of daylight that accompany the transition from season to season. A decrease in day length or in temperature might be sufficient to trigger the physiological responses that send the bird south; increases in those factors might trigger the converse physiological changes, sending the bird north again. The second general kind of explanation is that the impulse comes from within the bird.

In the past most research on the control of bird migration has concentrated on the influence of external factors. For the past 20 years, however, my colleagues and I have been studying the effect of internal rhythms on the migratory pattern. We have concluded that endogenous rhythms with a period of about a year have a powerful influence on when and how migration takes place. Such "circannual" cycles seem to provide an overall framework for the timing of migration. In addition they affect the details of the migratory flight, perhaps even helping the birds to navigate toward specific targets. Circannual rhythms can be modified by environmental factors and by learning. Nevertheless, in identifying them it seems we have discovered an essential component of the mechanism that controls avian migration.

Since our institute, the Vogelwarte Radolfzell, lies in southern Germany, it was natural for us to study species that participate in the Palaearctic-African system of bird migrations. The Palaearctic-African system is one of several great systems of avian migration, each of which encompasses the flight paths of many species. A second migratory system connects North America with Central and

South America; a third stretches from northeastern Asia to southeastern Asia and Australia. Since a large proportion of the birds that live in temperate areas migrate every year to southern regions, these migratory systems can be very large and complex. Of them, the Palaearctic-African one has been the most closely studied.

Much of what is known about the Palaearctic-African system has come from bird-banding experiments that began in Europe around the turn of the century. More than 60 million birds have been banded in Europe over the years, and the recovery of more than one million of them has yielded considerable information about their routes and wintering grounds. Partly on the basis of such information R. E. Moreau of the Edward Grey Institute in England has estimated that every year more than five billion birds invade Africa from Europe. The wintering grounds of the invaders extend from northern Africa to the Cape of Good Hope, and the distances they cover vary greatly. Whereas a woodcock may travel only a few hundred kilometers across the Mediterranean, a Siberian ruff may have to fly more than 12,000 kilometers across Asia and eastern Europe before reaching its winter home in central Africa.

For the most part our experiments have been carried out with European warblers (of the family Sylviidae) and flycatchers (of the family Muscicapidae). In studying how migration is controlled among birds of these species there were three main questions that had to be answered. The first was how the timing of migration is controlled. The second was how the bird is able to navigate to a specific target area on each leg of the migration. The third was how the organism can withstand the great energetic demands imposed by long-distance flying, particularly across oceans or deserts, where the bird cannot easily replenish depleted energy reserves.

All three questions are interesting, but investigating them simultaneously would have made for a somewhat unwieldy project. Accordingly our research strategy did not give the three questions equal attention. Instead, from the beginning of our work in the mid-1960's we concentrated on migrational timing. Specifically, we wanted to identify the source of the information that causes the bird to prepare for migration, begin its travels and ultimately terminate the migratory flight. In attempting to understand the problem of timing, however, we also obtained results that can aid in answering the other two questions.

Ultimately we carried out a long series of experiments aimed at elucidating the control of migratory timing. In many of them caged birds were carefully observed in order to detect the appearance of behaviors associated with migration. Such behaviors were of great significance in our work, because they served as indicators of the successive stages of the annual migratory cycle. One example is "migratory restlessness." The warblers and flycatchers we studied are nocturnal migrants: they fly at night and rest during the day. If the birds are held in cages during the fall or spring, when they would normally be migrating, they are intensely active at night. Such restlessness, which can take several forms, is a sign that the physiological events underlying migration have begun.

M igratory restlessness had a crucial role in the first experiment of the series, conducted in 1966. The goal of that work was to find out how the spring migration is initiated in the willow warbler (*Phylloscopus trochilus*). The willow warbler is a small, inconspicuous bird. Its breeding ground extends over much of central and northern Eurasia, and its wintering grounds are in tropical regions of the Southern Hemisphere. The willow warblers we worked with breed in Germany and spend the winter in equatorial Africa (see Figure 2.1).

When we began our work, it was already known that for some short-distance migrants changes in the photoperiod, or the average duration of daylight, provide an important stimulus for the beginning of migration. Of course in equatorial Africa, where the warbler spends the winter, the photoperiod undergoes little seasonal variation. Since the photoperiod varies so little, it seemed unlikely that light cues had much of a role in initiating the bird's spring migration. It was tempting to hypothesize that an internal timing mechanism triggers the homeward journey with no assistance from external stimuli.

To test that hypothesis I investigated the behavior of three groups of willow warblers. All three groups were held in experimental cages that made it possible to measure the degree of their migratory restlessness. In most warblers the nocturnal activity generally takes the form of hopping: the birds leap on and off their perches and jump around the cage almost continuously. Microswitches were mounted under one of the perches in the experimental cage. When the bird hopped on the perch, the micro-

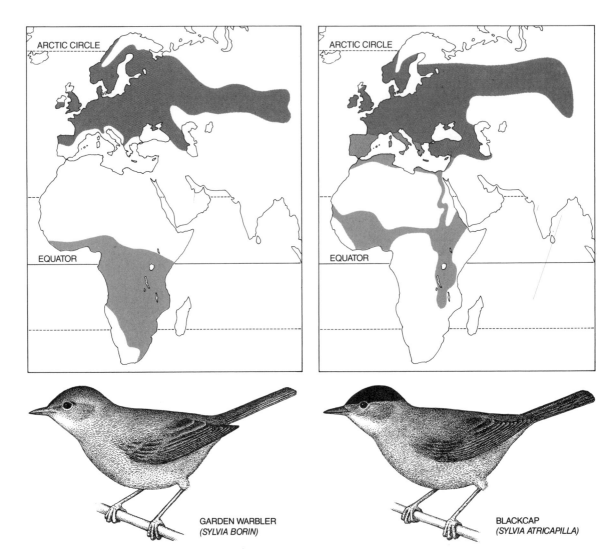

GARDEN WARBLER
(SYLVIA BORIN)

BLACKCAP
(SYLVIA ATRICAPILLA)

Figure 2.1 TWO SPECIES of warblers are among the birds studied by the author. At the left is the garden warbler and at the right is the blackcap (*Sylvia atricapilla*). Both are small and visually inconspicuous, but they are excellent songsters. Above each bird is a map depicting its breeding range (*gray*) and winter range (*color*). In fall the birds leave their breeding ranges and head south. Garden warblers spend the winter in central and southern Africa. Most blackcaps winter in southern Europe or in Africa north of the Equator; some blackcaps from eastern populations migrate to southern Africa. In spring the birds return to Europe and northern Asia to breed.

switch completed a circuit and caused a nearby event recorder to make a mark on a strip of paper. By counting the number of intervals (half hours, say) in which the bird was active, it was possible to measure the intensity of the restlessness.

One group of birds was held in our laboratory at a constant temperature and under a constant photoperiod of 12 hours per day. The aim of the experiment was to see whether those birds would undergo the seasonal event associated with migration on the normal schedule without external cues (see Figure 2.2). As a control, a second group was transported to a place in eastern Zaire at a latitude of about two degrees south that lies within the bird's normal wintering range. A third group was held throughout the winter at our institute in a room with large

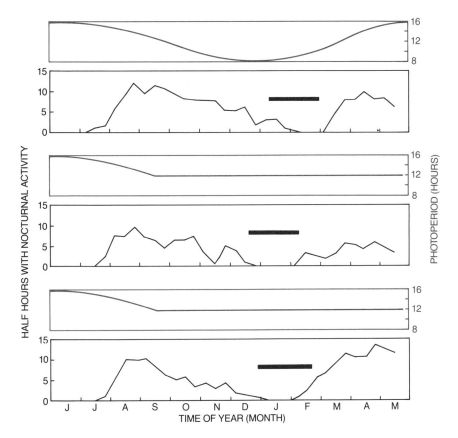

Figure 2.2 INTERNAL CUES underlie the timing of migration, as shown by three groups of willow warblers. Data for each group are presented in paired panels, depicting photoperiod (*shown in color*) and two of the birds' responses (*black*), namely migratory restlessness (*shown as a line*) and winter molt (*shown as a bar*). Warblers held captive in Germany (*top pair*) saw the normal variation in photoperiod there, while two other groups were denied seasonal stimuli. One group was taken to equatorial Zaire, where there is little variation in photoperiod (*middle pair*), and the other experience constant photoperiod in a closed lab. All responded on their normal yearly schedule.

windows and was thereby exposed to the variation in photoperiod typical of temperate regions. The three groups were held in individual cages to determine the end of the fall migratory restlessness and its onset in the spring.

The results of the experiment were quite clear-cut. In all three groups the fall migratory restlessness ended during December and January and the spring restlessness began during February and March. That is the pattern typical of willow warblers in the wild. Furthermore, each group molted (renewed its plumage) at the correct time. Molting patterns vary widely among bird species. For each species, however, the molt takes place at a fixed time in relation to the migratory flight. Free-living willow warblers carry out a full molt in midwinter, just after reaching their wintering grounds, and that is when all three groups of caged birds molted.

The fact that warblers under constant conditions of light and temperature adhere to the normal schedule of molting and restlessness suggested that the birds exploit internal cues to time their seasonal activities. The precise timing mechanism was unclear, however, because the results of the willow warbler experiment could be interpreted two ways. The correct timing of the events in winter might be due to an internal mechanism that is set in motion each spring in the breeding area. Such a mechanism would resemble an hourglass in that it would run for only a year before needing to be reset by external stimuli. On the other hand, the scheduling of seasonal events might reflect the operation of a mechanism that is more like a clock: an endogenous annual rhythm that continues to operate for many years without needing to be restarted.

The 1966 experiment did not enable us to decide which conclusion was correct, but my colleague Peter Berthold and I soon devised experiments that did. The bird we studied in the new experiments was the garden warbler (*Sylvia borin*), a species whose migratory patterns closely resemble those of the willow warbler. Three groups of garden warblers were exposed to a constant photoperiod for three years. For one group the period of light was 10 hours, for the second group it was 12 and for the third it was 16. In each case the results were the same: the migratory restlessness, molting and other related behaviors continued in a regular yearly pattern for the entire term of the experiment.

Subsequently we carried out an even longer experiment with garden warblers and blackcaps (*Sylvia atricapilla*). As before, the birds were exposed to an unchanging period of daylight. This time, however, we watched only for molting, which has turned out to be one of the most reliable indicators of cyclical yearly timing. Blackcaps and garden warblers in the wild molt twice a year. Under the experimental conditions both species continued to molt about twice a year for eight years, or even longer. The results of the two sets of experiments showed unambiguously that the small birds have a rhythmic internal timekeeper that functions to elicit the seasonal events in the right order (see Figure 2.3).

It was notable that under constant environmental conditions the period of the internal rhythm was not exactly a year. For example, the winter molts of

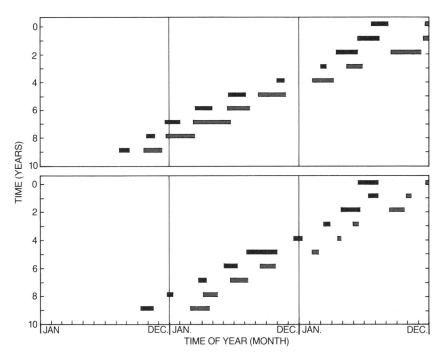

Figure 2.3 CIRCANNUAL RHYTHMS control the timing of seasonal events in some bird species, as shown in this decade-long study of birds that molt twice a year (winter and summer). When held under conditions of constant temperature and photoperiod, the molts of one blackcap (*upper panel*) and one garden warbler (*lower panel*) continued on a regular schedule. Black bars correspond to summer molts, colored bars to winter molts. Interestingly, the period between annual events (e.g., between consecutive winter molts) was only about 10 months; the birds molted somewhat earlier each year.

the caged blackcaps took place about every 10 months rather than every 12. This is the reason such endogenous oscillations are described as circannual, a word that comes from the Latin for "about a year." Circannual rhythms were first described some 25 years ago by Eric T. Pengelley and Kenneth C. Fisher of the University of Toronto in ground squirrels [see "Annual Biological Clocks," by Eric T. Pengelley and Sally J. Asmundson; SCIENTIFIC AMERICAN, April, 1971; Offprint 1219]. Since then similar periodic phenomena have been described in a broad range of organisms including coelenterates, insects, fishes, amphibians, reptiles and mammals in addition to birds.

The fact that the oscillations we observed in the garden warblers and blackcaps deviated from one year had two significant implications. First, it proved that the circannual rhythms could not have been due to the influence of uncontrolled seasonal stimuli present in the laboratory. If an uncontrolled seasonal stimulus had caused the observed rhythms, they would have had a period of precisely one year; that the circannual rhythms deviated from one year showed they are internal and spontaneous.

The second implication is that since in nature seasonal activities such as molting and migration always occur at the same time of year, there must be a factor that brings the approximately annual internal rhythms into correspondence with the solar year. The synchronizer was not difficult to find. We were able to synchronize a group of warblers so that they underwent migratory restlessness and molted on an exact annual cycle simply by simulating the yearly variation in photoperiod at 40 degrees north latitude, which lies in the warblers' European breeding ground. Even more striking was the finding that the circannual rhythm could be compressed to half its normal duration by increasing and decreasing the photoperiod on a six-month cycle instead of a 12-month cycle (see Figure 2.4).

So far we had established that the circannual rhythm, modified appropriately by environmental factors, sets the pace for the overall seasonal pattern of events. It seemed important to find out whether endogenous rhythms can also affect the details of the migratory flight, such as its duration and even its ultimate target. In order to study those questions we had to work with birds that we knew could not have learned a specific migratory route from experience.

Now, almost all the birds in our experiments are reared by human beings. When the birds are about a week old, they are separated from their parents and raised by the laboratory staff, who feed them worms, insect larvae and other tempting morsels. This process, which we call hand-raising, has at least two appreciable advantages for our kind of work. One is that by the time the birds are put in an experimental setup they are completely accustomed to the cage; hence they refrain from the erratic behavior seen in wild birds when first caged. Such erratic behavior would interfere with the experimental results, particularly in the study of migratory restlessness.

The other great advantage is that if the birds are hand-raised, their age, origin and history are known. As a result, when we came to study the influence of circannual rhythms on the details of the migratory pattern, we could rule out the influence of learning by working with birds we knew had never migrated. We found the inexperienced birds showed patterns of restlessness that mirrored the course of the migrational flight in wild birds. Caged willow warblers showed the most intense restlessness in August and September, which is the time free-living birds cross the Mediterranean Sea and the Sahara. Thereafter the nocturnal activity of the caged birds decreased, just as the migrational flight decelerates when the wild birds approach their winter home. The restlessness ended at about the time the free-living warblers reach the wintering ground.

Such results suggested that the willow warbler does not acquire the overall time course of its migration by learning. Additional work made it possible to generalize the conclusion. We found that in inexperienced, caged birds the duration of migratory restlessness was proportional to the distance covered by the species in its fall migratory flight. My colleagues Peter Berthold and Ulrich Querner showed that this correlation also holds within one species. They compared blackcaps from breeding grounds in Finland, Germany, France and the Canary Islands. The Finnish birds, which must travel farthest to reach central Africa, showed the most activity, followed respectively by the German, French, and Canary Islands birds, which travel progressively shorter distances.

Berthold and Querner exploited the variation among the subpopulations of blackcaps to show that the time course of migration is genetically programmed. Their strategy was to cross German blackcaps with those from the Canary Islands to see

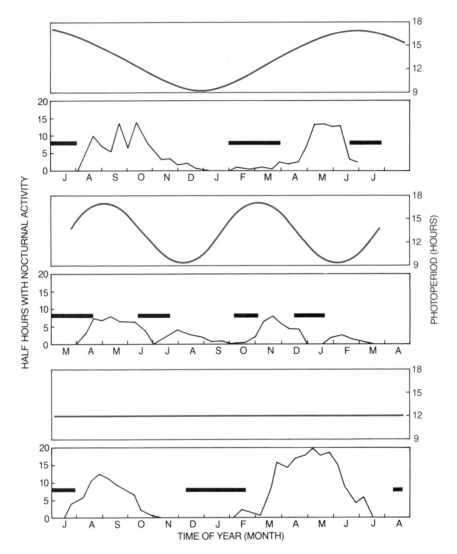

Figure 2.4 SYNCHRONIZATION OF CIRCANNUAL RHYTHMS is effected by external stimuli such as photoperiod. Because circannual rhythms have a period differing from one year, they must be synchronized so that the physiological events underlying migration coincide with the seasons. One group of garden warblers was exposed to the German photoperiod; they underwent migratory restlessness and molted on the normal, annual schedule (*top*). A second group, held under constant conditions, showed a circannual rhythm with a period of more than 12 months (*bottom*). A third group was exposed to a special six-month photoperiod cycle; their schedule for molting and migratory restlessness was compressed into six months (*middle*).

what chronological pattern of migratory restlessness the offspring would display. If, as they hypothesized, the pattern was genetic, then the offspring should show a degree of restlessness intermediate between the two parental groups. The reason is that any behavior as complex as migration probably re-

sults from the operation of many genes, and the inheritance of polygenic traits often takes the form of an average of parental extremes.

Breeding blackcaps in captivity is an arduous and frustrating task, but Berthold and Querner persisted until they had obtained 32 hybrids from German and Canary Islands parents. The birds were hand-raised and held in cages until their first autumn, when the intensity of their migratory restlessness was measured. The results were just as predicted by the polygenic hypothesis: the hybrid birds developed a pattern of nocturnal activity almost exactly intermediate between those of the parental populations.

The breeding experiment carried out by Berthold and Querner showed indisputably that the temporal course of the migratory restlessness is determined by a genetic program. It soon became clear that this temporal program may also help young birds to find the wintering ground, which is clearly essential for survival.

Among species that migrate in groups, inexperienced migrants can follow more experienced leaders. Many species, however, migrate singly, including most of the birds we work with. Among those species each bird must be able to find the wintering grounds on its own, whether it has ever made the journey before or not. Clearly, if a bird has not migrated before, learning cannot help it to find its way, and so some other mechanism must serve for navigation. Evidence is accumulating that a temporal program is one such mechanism.

It has been known for some time that novices rely on methods of navigation different from those of experienced migrants. This was made strikingly clear in work done by A. C. Perdeck of the Institute for Ecological Research in the Netherlands. He studied European starlings, which breed on the shores of the Baltic Sea in Sweden, Finland, Latvia, Lithuania, Poland, Germany and Denmark. In the fall the starlings fly southwest to wintering grounds that are mainly in northern France and southern England. For one experiment Perdeck trapped 11,000 starlings near The Hague and had them transported by airplane to Switzerland, where they were banded and released.

More than 300 of the birds were later recovered, and when the recovery points were plotted on a map, an intriguing distribution emerged (see Fig 2.5). Birds that had migrated at least once before were able to compensate for the southward displacement. They were found scattered along a flight path leading toward the wintering grounds, a path that entailed making a sharp northward turn from the original course. The young birds, on the other hand, were found along a flight path more or less parallel to the normal one, leaving them hopelessly far south of the usual target. One must conclude that the inexperienced birds were unable to compensate for the displacement and had simply kept flying in the original direction after being released.

Perdeck's result suggests that young birds can fly in only one general direction when migrating. A navigation system capable of accommodating only one flight heading is primitive, but it is by no means completely impractical. If the initial heading is correct and the external perturbations (such as those caused by wind) are not too great, all the flier needs to reach the target is a clock that indicates when to start flying and when to stop.

Various field experiments have yielded results consistent with the hypothesis that inexperienced birds depend on a navigational clock. Perhaps the clearest demonstration was made by W. Rüppell and E. Schüz of the Vogelwarte Rossitten, the predecessor of our institute. They trapped young carrion crows about halfway along their migratory route from Baltic breeding areas to wintering areas in northern Germany. The birds were banded and taken to a location beyond the normal winter range before being released. When some of the banded crows were recaptured during the winter, it was found they had continued to migrate on the original heading, which now took them away from the wintering ground. Moreover, they had flown a distance roughly equal to the distance that separated them from the target when they were captured. It would seem that the migratory clock had simply continued to tick after the displacement, with the result that the young crows had flown the correct distance, but away from the target.

Perdeck's results, combined with those of Rüppell and Schüz, help to clarify the difference between the navigation systems of experienced and inexperienced birds. Birds that have migrated before have the capacity for true navigation: when displaced, they can correct their course and find the target. (How they accomplish the feat is not yet understood.) Inexperienced birds have no such capacity; they apparently rely solely on a single heading and a preset flying clock. Until recently the nature of the navigational timekeeper was not known. Our work with warblers, however, suggests the clock is provided by the genetic program that specifies the temporal course of migratory restlessness. The physiological rhythm underlying the restless behavior seen in the caged birds tells the mi-

Figure 2.5 NAVIGATION SYSTEM of inexperienced birds differs from the one used by adults that have made the migratory flight at least once before. A. C. Perdeck captured starlings about halfway along their fall migratory route (*gray*) from Baltic breeding areas to wintering grounds in France and England. The birds were transported to Switzerland and released. Experienced migrants (*color*) compensated for the displacement and flew toward the normal winter range. Inexperienced birds (*black*) kept flying in the original direction, on a course that took them toward the Iberian peninsula. It seems inexperienced birds are not capable of true navigation; they can fly only in one fixed compass direction.

grant when to take off, when to expend peak effort, when to decelerate and when to stop flying.

Although their navigation systems differ considerably, novices and experienced birds share the need to know the correct flight heading. Intriguingly, it appears circannual rhythms can help to solve that problem too. For some time experimental results have been available implying that the migrant's choice of a compass heading is determined by its physiological state rather than by external cues such as geographic information or the position of the stars. Among the data are some obtained by Stephen T. Emlen of Cornell University [see "The Stellar-Orientation System of a Migratory Bird," by Stephen T. Emlen; SCIENTIFIC AMERICAN, August, 1975].

Emlen exposed a group of indigo buntings to two photoperiodic cycles a year. The accelerated changes in photoperiod advanced the timing of molt and of migratory restlessness in the experimental birds. Eventually the birds came into an "autumnal" physiological state in the spring. They were then put in orientation cages in a planetarium under a simulated spring sky. In spite of the celestial

cues, the birds responded to their internal state and consistently hopped southward, as if attempting to begin the fall migratory flight.

From the point of view of the work my colleagues and I have been doing, Emlen's results could be interpreted in two ways. The buntings' choice of the incorrect direction could be due either to the effect of the photoperiod or to that of the underlying circannual rhythm. We needed to know which factor was responsible. Therefore, with Wolfgang Wiltschko of the University of Frankfurt as my collaborator, I set out to design experiments that would disentangle the two effects. Wiltschko and I did our experiments on 59 hand-raised garden warblers (see Figure 2.6). Garden warblers are among a large group of European birds that share a basic migratory route. They leave their breeding ground in the fall on a southwesterly course. Over the Iberian peninsula or northern Africa, however, they turn south or south-southeast and continue on that heading until they reach their winter range. The return flight in spring has a more direct route: the birds fly straight from south to north.

As in our previous work, to isolate the effect of

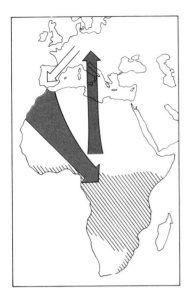
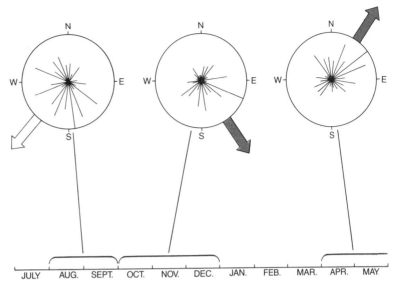

Figure 2.6 DIRECTIONAL PREFERENCES of garden warblers correspond with the circannual rhythms. These birds' migration route is triangular (*map at left*). In September they fly southwest to Spain, then head southeast to equatorial Africa; in the spring they go due north. The three circles at the right show their directional preferences at different times of year, when held under constant conditions of light and temperature in orientation cages (see Figure 2.7). The lines within each circle show the birds' preferences, with the large arrow projecting from the circle showing the overall choice. These results match well with the seasonal migratory directions of free-living birds.

the circannual rhythm we needed to minimize the seasonal fluctuations that the birds experienced. The warblers were held in chambers with constant temperature and an unvarying photoperiod.

In order to test a bird's choice of direction, the bird must be able to orient itself in relation to the points of the compass. Wiltschko had shown previously that the garden warbler can orient itself by means of either the stars or the earth's magnetic field. The celestial map changes with the season. Since we intended to eliminate all seasonal cues, the birds were not allowed to see the sky. The earth's magnetic field, on the other hand, undergoes very little seasonal variation. We made no effort to insulate the birds from the magnetic field, and they were thereby able to orient themselves.

The experimental system was nearly complete. The warblers had been isolated from seasonal cues and provided with a means of sensing compass directions. Now we needed a way to test the caged birds' directional preferences during the migratory period. In order to observe their directional preferences the warblers were put in special cages at regular intervals throughout the year. Each cage was an octagon with a diameter of about a meter. Eight perches radiated from the center of the cage like the spokes of a wheel. Microswitches mounted under the perches made it possible for us to tell which perches the birds preferred during the night in each phase of the migratory cycle (see Figure 2.7).

The results were quite dramatic. In August and September, when wild garden warblers head southwest on the first leg of their triangular flight path, the caged birds tended to hop onto the perches at the southwest corner of the cage. During the second half of the fall migratory season, when the migrating birds turn south and head for central Africa, the caged birds preferred to perch in the south and southeast parts of the cage. In the spring, when the migrants head straight for Europe from their equatorial winter homes, the caged birds most often chose the north perches.

It seems clear that circannual rhythms can control the direction of migration as well as its initiation and its temporal structure. Indeed, the overall result of the work done in my laboratory in the past two decades has been to emphasize the effects of the spontaneous internal clock. Yet it should not be supposed that the operation of the circannual

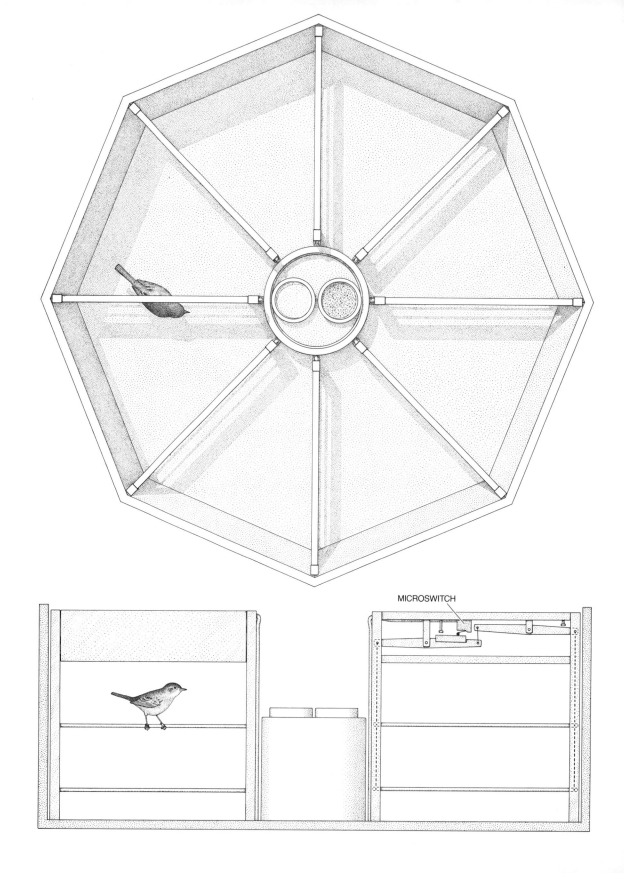

MICROSWITCH

rhythms eliminates the influence of external cues. As I have described above, circannual rhythms can be modified by changes in photoperiod or—in adult birds—by navigational learning. Some recent work has been aimed at finding out more about modifications of the internal program; my colleague Herbert Biebach has shown that the modifications can be quite specific.

Biebach concentrated on the changes in weight that birds undergo while they are migrating. Before taking off, the migrants generally eat more intensively than usual in an attempt to acquire stores of fat to be used as fuel on the journey. The weight gain can be considerable. Garden warblers in the field can grow from an initial weight of 16 grams to a final weight of 30 grams, and in the laboratory (where an unlimited supply of food is available) the gain can be even greater. The accumulated fat stores dwindle rapidly as the birds make the exhausting migratory flight.

Biebach trapped migrating songbirds at two sites in the Sahara. One site was an artificial resting area made of a few bushes planted in the sand. Birds were attracted by the bushes and touched down, but no food or water was available there. The other site was an oasis with abundant vegetation and a rich insect life. At both sites many of the birds present in the morning were trapped, banded, weighed and then released. Some birds were later recaptured and weighed again. It turned out that the birds behaved quite differently at the two sites. At the desert trap, where there was no food, all the birds left on the evening of the day they arrived. At the oasis, however, a fairly large proportion of the arriving birds stayed for more than one day. The weight data showed that the birds remaining for more than a day tended to be the ones whose fat stores had been depleted on arrival.

Biebach's findings suggested that the combination of low fat reserves and the opportunity to

replenish them inhibits the endogenous migratory program, which would otherwise impel the birds to continue their flight. That conclusion has since been confirmed in the laboratory. The laborataory studies show that neither the depletion of the fat reserves nor the opportunity to feed is on its own sufficient to interrupt the operation of the program; both factors must be present simultaneously for the migratory impulse to be inhibited. As soon as the bird attains a certain minimum body weight the impulse reappears.

The capacity of the endogenous program to be overridden by external stimuli has decided advantages. If the circannual rhythms could not be modified, a bird whose fat reserves were depleted might starve to death while being driven on by the relentless migratory impulse. If the impulse can be temporarily overridden, the migrating bird has a chance to replenish its stores of fat before resuming its flight. In one sense the behavior Biebach has illuminated so nicely is similar to the navigation system of the experienced birds. In both instances a rather inflexible mechanism is overridden by a system that can adjust the internal time program to specific external conditions.

One of the main tasks facing students of bird migration is to find out just how the endogenous program interacts with external stimuli to yield the observed migratory behavior. Another problem, equally fundamental, is to identify the physiological basis of the circannual rhythms. My colleagues and I are currently working on both questions, but it may take many years to answer them. After all, it has taken two full decades to achieve a detailed understanding of how the circannual rhythms operate. In those two decades it has become clear that endogenous rhythms provide the overall framework of migratory behavior. The details of the migratory cycle can be modified in a variety of ways to accommodate external circumstances, but the impulse that drives the great V of geese as well as the ragtag group of starlings comes from yearly rhythms originating deep in the organism.

Figure 2.7 ORIENTATION CAGE for testing directional preferences of migratory birds has eight pairs of perches (*upper panel*). Each perch is connected by a string to a microswitch (*lower panel*). When the bird jumps on one of the perches, the microswitch completes a circuit and a nearby event recorder registers the electrical impulses. The apparatus makes it possible to find out which part of the cage (and thus which compass direction) the bird prefers. By this means the author was able to show that the directional preferences of migratory birds undergo seasonal changes as the result of endogenous rhythms.

Memory in Food-hoarding Birds

Birds that hide seeds and later recover them appear to excel in spatial memory. One species evidently remembers where it has put thousands of caches for as long as several months.

. . .

Sara J. Shettleworth
March, 1983

In England in the winter a bird feeder stocked with peanuts is all the apparatus one needs to get an animal to show behavior suggesting it has remarkable powers of memory. Indeed, its memory appears to far surpass in capacity the memory that any other animal has ever shown in the laboratory. The feeder attracts many birds. Among them are great tits, blue tits and marsh tits, which are all small, lively birds related to the North American chickadees. The great tits and the blue tits congregate at the feeder, eating as fast as they can. They interrupt their meal only to chase away their competitors. A marsh tit nonetheless darts in, grabs a peanut and flies off. It is back almost immediately to grab another. It stores the peanuts nearby, each in a different site, until the feeder is empty. Then it searches out its hidden food.

In the American Southwest another bird, Clark's nutcracker, a relative of the jays and crows, shows similar behavior. In the late summer the nutcracker harvests the seeds of piñon pines. It repeatedly fills its sublingual pouch, a pocket under the tongue, and then flies several miles to bury the seeds. Often the burial sites are on bare, south-facing slopes where the snow will not be deep later on. A nutcracker may bury as many as 33,000 piñon-pine seeds in caches of four or five seeds each (see Figure 3.1). Throughout the winter it returns and digs up its thousands of caches.

How do these birds find their hoards? Does that tit or the nutcracker remember where it has stored each peanut or cache of piñon seeds? Until recently observers of hoarding behavior in the wild doubted that the birds rely on memory. For one thing the bird would need a capacious memory indeed to remember the sites of hundreds or thousands of individually hoarded items. The memory would also have to be long-lasting. Even a short-term hoarder such as the marsh tit does not recover its stores until hours or days after it deposits them, and a long-term hoarder such as Clark's nutcracker does not return to a hoard for months, perhaps not until spring, when nutcrackers feed their young on piñon seeds. Moreover, in principle the hoards could be recovered without the aid of memory. A hoarder could store food only in certain kinds of sites, such as south-facing slopes or holes in bark. To recover its stores it would need only to search in sites of that kind. Then too it could employ memory only for the area in which it hoarded rather than for the individual sites. Inside that area it could search by trial and error or conceivably by cues such as odor.

Figure 3.1 WHEN HOARDING FOOD a Clark's nutcracker jabs its bill into the ground to loosen the soil (*a*); then it inserts a conifer seed (*b*), rakes soil and grass over the seed (*c*) and places a pebble on top of the cache (*d*). The pebble apparently serves only to camouflage the site until weather obliterates all traces of the caching. The procedure takes from 10 to 20 seconds. In nature the nutcracker hoards as many as 33,000 piñon-pine seeds in thousands of widely distributed caches that average four or five seeds each. The bird does its hoarding from September through November and relies on memory to find the caches throughout the following winter, spring and summer.

Psychologists studying learning and memory in animals are becoming increasingly aware, however, that certain species have adaptive specializations that make them particularly good at learning and remembering things it is important for them to know. Among the well-known instances are the ability of many birds to learn their species' songs, the ability of rats and other animals to remember spatial locations, the ability of bees to remember the location of flowers and the ability of many animals to learn to avoid noxious food. If food-storing birds really do remember large numbers of storage sites over long periods, their memory could be another example of an adaptive specialization, one that would enable the birds to recover their stores efficiently. A bird that could remember where it had hidden its food would make fewer errors recovering it and spend less time and energy than birds searching randomly. Recent studies indicate that at least some food-storing birds do remember the sites of their hoards quite well.

Some of the most detailed evidence that food-storing birds remember their hoards has been accumulated over the past few years by John R. Krebs and his associates at the University of Oxford. They began by looking for evidence confirming that marsh tits in their natural environment recover their hoards. To that end marsh tits were trained to come to sunflower-seed dispensers set up in their territories in Wytham Wood near Oxford. Then Richard J. Cowie, Krebs and David F. Sherry coated the husks of the seeds with a radioactive substance harmless to the birds. (The birds remove the husks before eating the seeds.) The radioactivity enabled the investigators, equipped with a portable scintillation counter, to find the seeds the birds had cached and then check every few hours to see when they disappeared (see Figure 3.2).

The seeds of course might be eaten by other birds or by rodents such as mice and voles. Cowie, Krebs and Sherry reasoned, however, that if marsh tits use their memory to recover their own stores, the seeds should disappear sooner from the birds' storage sites than they would from false sites the investigators established. This was indeed the case. The seeds the birds stored were depleted sooner than seeds the investigators placed in similar sites that were each one meter from a hoard made by a bird. In Wytham Wood in the winter the natural hoards were generally gone in a day or two. The last of the false hoards disappeared fairly soon afterward, suggesting that predation of the hoards is common.

One defensive tactic the marsh tits apparently adopt against animals that steal from their hoards is to disperse seeds rather than burying them in clusters. On the average the hoards the marsh tits established in Wytham Wood were seven meters apart. Sherry, Mark Avery and Allen Stevens set up false hoards of sunflower seeds and found that the seeds stayed in place longer the farther apart they were, up to a distance of about seven meters. Hoards farther apart than that were no safer than the ones at seven meters. In Wytham Wood, at least, marsh tits seem to have hit on the optimal spacing.

Having obtained evidence that marsh tits in nature recover their hoards, Krebs and his colleagues turned to observations in the laboratory. Marsh tits supplied with a bowl of seeds and with suitable storage sites hoard quite readily in captivity. Sherry, Krebs and Cowie offered marsh tits trays of moss one meter on a side in which to store sunflower seeds. The birds did appear to remember where they had buried seeds even after 24 hours. In these experiments, however, the behavior of the birds was recorded only as visits to relatively large areas on the tray. Moreover, in an effort to ensure that the birds employed only memory in the experiments, the hoarded seeds themselves were removed from the tray before each bird was allowed to return.

Thus it was difficult to say with what precision the birds located their hoarding sites. It was equally difficult to identify mistakes the birds might be making.

When a Guggenheim Fellowship enabled me to spend a sabbatical leave at Oxford, my interest in adaptive specializations of learning and memory attracted me to the marsh-tit studies. Krebs and I designed some experiments to analyze in more detail the marsh tits' apparent memory. We were particularly interested in finding out whether the items most recently stored by the birds are the ones they recover first. Such a "recency effect" often appears in tests of memory, and in the case of food hoarding there is good reason to expect it: if hoards are lost to scavengers at a constant rate, the most recent hoards will be the ones most likely to be available to the hoarders at any given time. Looking for them first is therefore the hoarders' best bet.

In a large room Krebs and I set up sections of tree branches. In each section we drilled a number of holes the right size for storing hemp seeds. There were about 100 holes in all. Each hole was covered with a flap of cloth, which a bird would have to lift in order to store a seed or look for one.

The room was the site of a series of experiments. For each trial of our first experiment we allowed a marsh tit to store 12 seeds it got from a bowl on the floor in the middle of the room. Each hole could hold only one seed; thus each seed was stored in a different hole. Once the seeds were cached we kept the bird outside the room for about two and a half hours. We removed the bowl of seeds; then we readmitted the bird, and it searched for the seeds it had stored. If it searched at random among the 100 holes, it would have to investigate an average of about eight holes to find one seed. The birds were much more efficient (see Figure 3.3). On the average each bird made about two errors per seed. At the beginning of a recovery test a bird sometimes went to three or four seeds in succession without looking in empty holes.

Maybe instead of employing memory the tits were smelling the seeds or detecting them in some way we had not thought of. Smell seemed unlikely; the sense of smell is generally poor in birds. Still, to test these possibilities we allowed a marsh tit to store 12 seeds and then we moved the seeds to other holes (ones the bird itself had used in earlier trials). Under these circumstances the readmitted birds made about six errors per seed. They eventu-

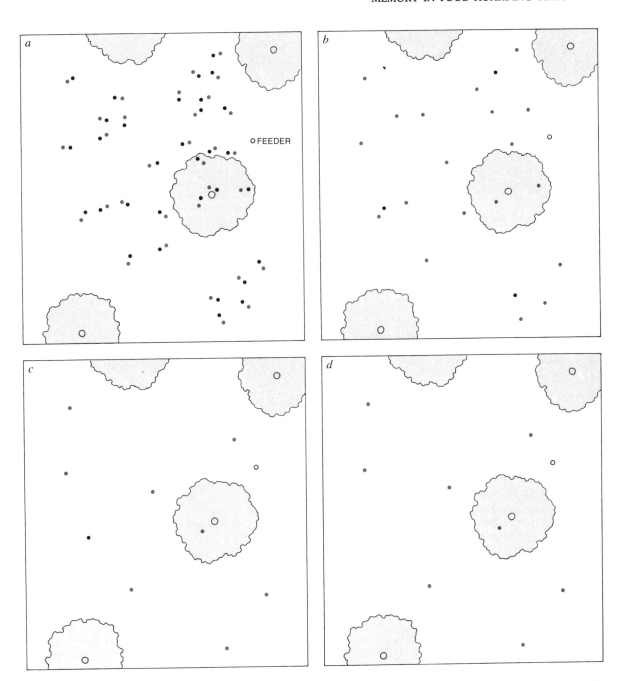

Figure 3.2 DISAPPEARANCE OF CACHES from Wytham Wood near Oxford was the first evidence confirming that marsh tits use memory to recover hoarded food. Richard J. Cowie, John R. Krebs and David F. Sherry of the University of Oxford coated 30 sunflower seeds with a radioactive substance, then employed a portable scintillation counter to find where a marsh tit had cached them (*black dots*) in Wytham Wood (*a*). One meter from each cache the investigators established a false cache (*colored dots*). Ten hours later (*b*) 19 false caches were undisturbed but only three of the bird's caches remained. The investigators made further checks after 19 hours (*c*) and after 30 hours (*d*).

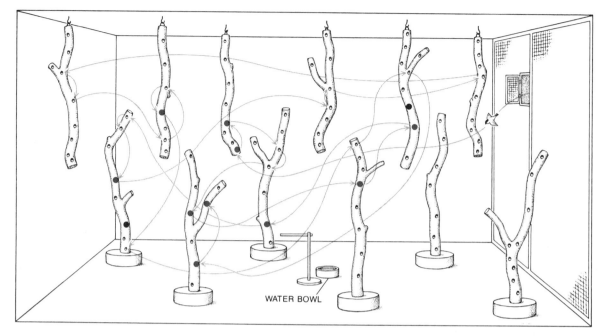

WATER BOWL

Figure 3.3 FOUR EXPERIMENTS testing the memory of marsh tits were done by the author in collaboration with Krebs. In each instance a bird stored seeds in some of the roughly 100 holes the experimenters had drilled in tree branches in the laboratory. Here a marsh tit hoarded 12 seeds (*dots*). Readmitted to the room two and a half hours later, it found 10 of the seeds (*colored dots*) by inspecting only 24 holes.

ally found about two-thirds of the seeds by looking in more holes than usual (see Figure 3.4).

The results indicate that the marsh tits did not detect the stored seeds by smell or some similar cue. On the other hand, the performance of the tits in the first experiment does not necessarily mean that they remember storing each seed in a certain hole out of the 100. To some extent each bird in the laboratory grew to have favorite holes in which it stored seeds particularly often. Each bird was likely to inspect its favorite holes when it was searching for stored seeds. The recovery of seeds from such holes could be due to nothing more than the bird's habit of visiting those places both when it stores seeds and when it recovers them. Indeed, this tendency probably explains why the birds' performance was better than chance when we moved the stored seeds.

A closer look at our data showed that a tit did not simply go to the same holes on each test. Even the holes the tit preferred most were more likely to be inspected during the recovery of seeds if the tit had actually put a seed there. Nevertheless, the problem

was fundamental. In conventional laboratory tests of memory the experimenter supplies the information the animal must try to remember. In our tests the birds themselves produced the information when they stored seeds where they chose. Any tendency they might have to look into the same holes on successive visits to the test room would lead them to perform as if their memory for the storage sites were good.

Our next experiment was therefore designed to make the birds' memory work against their tendency to prefer certain storage sites. Again each bird stored seeds, was removed from the room and then readmitted some two hours later. Instead of requiring the bird to search for the seeds during the second session, however, we now allowed it to store additional seeds. To get it to do so we simply left a bowl full of seeds in the room. We reasoned that if the tits remembered which holes already had seeds, they would avoid those holes when they stored a second batch. Conversely, if they inspected the same holes each time they visited the room, many

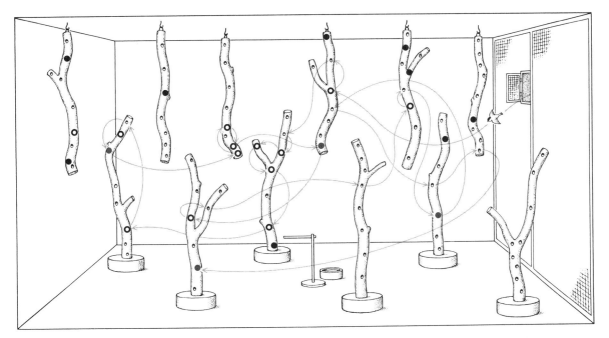

Figure 3.4 IN A SECOND EXPERIMENT a marsh tit stored 13 seeds (*open black circles*), but Krebs and the author then moved them to other holes (*dots*). Among the first 24 holes the bird inspected when it was returned to the room, 11 were holes it had employed. The experiment indicates that the bird was not detecting the seeds by a cue such as their odor. The bird did succeed in finding four seeds (*colored dots*).

of the holes they inspected on their second visit would already have a seed in them.

The results were clear. The birds almost never inspected holes containing a seed when they were storing a second batch of seeds (see Figure 3.5). When they were hungry, however, and there were no more seeds to store, they showed their usual efficiency in recovering seeds from holes. We were interested to note that a bird storing a second batch of seeds often chose a hole near one it had employed in storing the first batch. Thus the bird seemed to remember and avoid the individual holes, not just particular branches or parts of the room.

After a bird had stored two batches of seeds, we could let it recover all the seeds (see Figure 3.6). In this way we could test for the recency effect. If marsh tits show the effect, they should recover seeds from the second batch sooner or more often than seeds from the first batch. Our results suggest that there is a recency effect but that it is not strong. If the memory of storage sites fades slowly, how-

ever, the two episodes of hoarding would have to be separated by more than two hours for the recency effect to be clear.

An additional finding emerged from the experiments. It is plain that an efficient searcher for stored food should seldom inspect empty sites. Furthermore, it should be able to remember where it has already been to recover food so that it can avoid going back. Laboratory rats faced with the task of collecting food from a number of sites in a large maze prove to have the latter ability. Marsh tits have it too. In the experiments that Krebs and I did, the holes a bird inspected as it recovered seeds were inspected a second time much less than one would expect if the bird were not remembering and avoiding such sites (see Figure 3.7). In experiments done by Sherry, birds tended not to search areas of moss they had already searched for seeds. A marsh tit searching for stored food makes use, then, of two kinds of information: it remembers where food has been stored and also the sites it has already inspected.

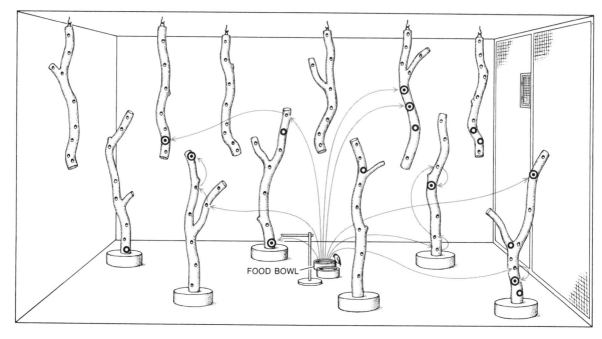

FOOD BOWL

Figure 3.5 TWO BATCHES OF SEEDS were hoarded by a marsh tit in the third experiment. In choosing holes for the second batch (*black circles with dots*) it avoided the holes in which it had stored the first batch (*open black circles*). The experiment establishes that the bird was not merely visiting a particular set of holes whether it was storing seeds or recovering them. The marsh tit could remember where it had been.

If information about hoarding sites is stored in the marsh tit's brain, the bird's memory should be subject to a peculiarity of the nervous system that seems to be present in many birds. In many vertebrate animals (including man) information about the left half of the visual field comes from both eyes, but it goes to the right half of the brain, and vice versa. The situation is somewhat different in birds such as the marsh tit, where the eyes are on the sides of the head and each eye surveys a separate visual field. The tracing of information pathways in the brain of the bird and the results of experiments done mostly with pigeons suggest that in these birds information reaching the brain by way of one eye is stored primarily in the half of the brain on the opposite side of the head. In short, such birds have little or no interocular transfer for many kinds of information. Sherry, working with Krebs and Cowie, employed this fact to interpret an intriguing behavior marsh tits show when they store a seed. Having tapped or poked the seed into place, the bird quickly cocks its head first to one side and then to the other. It is as if the bird were looking at the storage site or at the landmarks near the site with each eye in turn.

Does the marsh tit cock its head in order to store visual information about hoarding sites in both halves of the brain? To answer this question Sherry had marsh tits store sunflower seeds wherever they chose in a tray of moss while each bird was wearing a translucent cover over one eye. If the bird was later allowed to search for the seeds with the same eye covered, it found the seeds quite normally. If the patch was moved to the other eye, however, so that the bird had to search with the eye that had not viewed the storage sites, the bird behaved as if it did not remember where the seeds were. If marsh tits do have little or no interocular transfer of information, the behavior is further evidence that they rely on their memory to recover their stores.

What about other food-storing birds? In the past few years several investigators have been studying birds such as Clark's nutcracker, which makes thousands of winter caches. Again evidence of memory has been sought both in the field and in the laboratory. Of course a field study attempting to discover

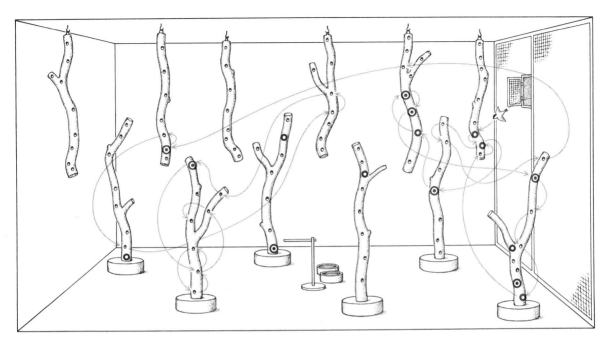

Figure 3.6 RECOVERY OF BOTH BATCHES of seeds by the marsh tit that had hoarded them was the final experiment in the series. By inspecting 29 holes it found 10 seeds: four (*open colored circles*) from the first batch of eight and six (*colored circles with dots*) from the second batch of eight. Its first seven inspections of holes were all successes. In two instances the bird made the mistake of revisiting a hole.

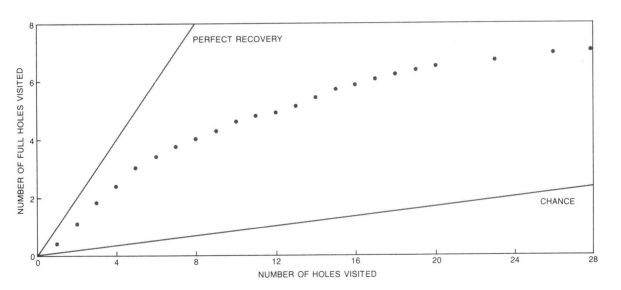

Figure 3.7 MARSH TIT'S PERFORMANCE was charted by Krebs and the author. If the bird were searching at random for eight seeds in 100 holes, it would inspect about 12 holes to find each seed. If its memory were perfect, on the other hand, it would inspect just one hole to find each seed. The bird's performance was intermediate: in 12 trials it inspected an average of about 3.7 holes to find each seed.

such things as whether or not a Clark's nutcracker recovers its own caches is extremely difficult to do because of the large number of caches and the long interval between storage and recovery. Few people would want to wait several months near a nutcracker's cache, hoping to learn whether it was eventually dug up by the bird that buried it.

Diana Tomback of the University of Colorado at Denver reasoned, however, that much could be learned from the way Clark's nutcrackers search for caches. She took advantage of the fact that the birds leave a record of their searches in the form of beak marks in the snow and the earth where they dig. Moreover, successful searches can be identified by the presence of piñon-seed coats next to the holes from which seeds were taken. If nutcrackers search at random, successful probe holes and clusters of unsuccessful holes should be more or less evenly distributed over landscapes where the birds have searched. Tomback found, on the contrary, that unsuccessful probes were clumped around successful ones. The pattern indicated the nutcrackers were not searching for caches by trial and error. Furthermore, in the early spring (before rodents had found many of the caches) about two-thirds of the probes were successful, far more than one would expect of a random search.

Tomback's observations do not prove unequivocally that the birds could remember where their caches were. They might have smelled them (although again it is an unlikely possibility). Alternatively, they could have searched mainly in places of the kind that are likely to contain caches. Recent laboratory studies done by Stephen B. Vander Wall at Utah State University seem to rule out these possibilities. Vander Wall studied four captive Clark's nutcrackers in a large outdoor aviary. Two of the birds readily cached seeds in the sandy floor of the enclosure; the other two did not. Nevertheless, the nonhoarders did dig in the sand with their beaks and eat the buried seeds they happened to uncover. Vander Wall allowed the two hoarders to bury seeds in the aviary; then he allowed all four birds to search for seeds.

The results provided dramatic evidence of memory. When the nutcrackers that had hoarded seeds dug in the sand to find them, 70 percent of their probes were successful. When the nonhoarders dug, they found seeds in only 10 percent of their probes. To be sure, a success rate of 10 percent is substantially greater than one would expect for a bird probing completely at random. On the other hand, the two nutcrackers that cached seeds preferred to make their caches near prominent objects in the aviary, such as logs and rocks, and all four birds searched most often near those objects. This tendency accounts for the nonhoarders' success.

Perhaps nearby prominent objects serve as cues for memory when the birds attempt to find their stores. To test this hypothesis Vander Wall covered the aviary floor with a plastic sheet, leaving exposed only an oval area where the sheet had been cut away. At each end of the oval he arranged four large objects such as rocks. The nutcrackers were allowed to make caches wherever they chose in the uncovered oval (see Figure 3.8). Then one end of the oval was extended by 20 centimeters and the four objects at that end were moved in the direction of the lengthening so that they too were each 20 centimeters from their original position.

Would the nutcrackers be able to find caches at the altered end of the oval as well as the ones at the unaltered end, or would their probes be off by 20 centimeters? The latter proved to be the case. Toward the unaltered end of the oval the birds' success rate was high. Toward the altered end the success rate declined. Most of the probes at the altered end were within a few centimeters of the point that was "correct" with respect to the altered position of the nearest large object. The birds also made errors near the middle of the oval, but the errors were less extreme. This finding suggests that the cache sites had been established with reference to both sets of objects.

For any species that disperses its stores and relies on memory to recover them one may ask whether the animals' memory enables them to benefit in other ways. That is, are food-storing species better in general at remembering where things are? An attempt to answer this question would require that animals be given tests of spatial memory that do not involve stored food. Such tests have not yet been done with food-storing birds, but thinking about how they might be done reveals some important points.

Among the laboratory techniques employed by psychologists who investigate memory in animals the closest analogue to the hoarding of food is a delayed-response experiment. Here an animal is shown some visual cue, such as a flashing light or a geometric figure. Later it is offered a choice between the original cue and another one. It is rewarded for

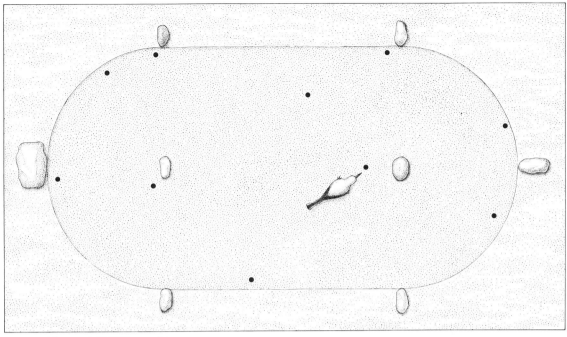

|← — .5 METER —→|

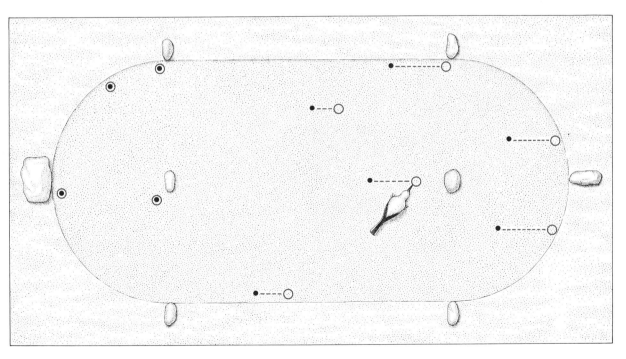

Figure 3.8 PROOF OF MEMORY in Clark's nutcrackers emerged from experiments by Vander Wall. A nutcracker was allowed to cache seeds (*black dots*) within an oval sandy area; four stones were arrayed at each end of the oval (*top drawing*). The oval was then lengthened by 20 centimeters and the stones at the right were moved accordingly (*bottom drawing*). Two days later, when the bird was allowed to search for its caches, it dug successfully at the left (*colored circles*). At the right, however, it made errors of approximately 20 centimeters. Evidently it was relying on its memory of the positions of nearby stones. Errors in the middle suggest the bird estimated the position of some caches with reference to both sets of landmarks.

choosing the one it saw before. The performance of a laboratory rat or a pigeon typically falls to the level of chance if a few seconds or minutes pass between the first viewing and the second.

Recently Donald M. Wilkie and Russel Summers of the University of British Columbia trained pigeons in a variant of the delayed-response procedure that requires spatial memory. The work illustrates some of the differences between food hoarding and conventional laboratory studies of animal memory. In Wilkie and Summers' experiment a pigeon faced a three-by-three array of nine white disks that could be lighted individually from behind. At the start of each trial one disk was lighted briefly. A few seconds later the disk was lighted again, along with another disk or several others. The pigeon was rewarded (it received a pellet of food) if it pecked the disk that had lighted up twice.

The pigeon's task seems to be quite similar to the task facing a food-storing bird. After all, a bird storing food is exposed to visual cues at the storage site, and in order to recover the food it must select the site in preference to similar sites nearby. Yet with only one cue to remember at a time, the pigeons in Wilkie and Summers' experiment appeared not to know which disk had been lighted only a few seconds after they saw it. In contrast, a Clark's nutcracker evidently remembers thousands of storage sites for months.

The pigeon's task did differ in a number of ways from the nutcracker's. The pigeon passively viewed the cue it would be called on to remember. The cues themselves were invariably within a few centimeters of one another and differed only in their position within the array. The food with which the pigeon was rewarded was not behind the cue itself. Any of these factors would be expected to make the pigeon's task more difficult than the task facing a food-hoarding bird. For instance, laboratory investigations done mainly with rats show that animals can remember for several hours the locations they actually visit.

Perhaps, therefore, the types of cues available at hoarding sites are what make the sites memorable. On that hypothesis a food-hoarding bird would not necessarily have a memory extraordinarily better than that of other animals. By adopting a life of hoarding food it would simply have put itself in circumstances where its memory is particularly serviceable. How could this be tested? Comparisons of memory between species that hoard food and species that do not are often compromised by differences in the motivations and the sensory and motor abilities of the species. Obviously the comparisons would be most significant if the species were closely related ones.

In this respect it is notable that among the corvids (the jays, crows and nutcrackers) and also among the tits some species store food whereas others do not. Among the tits, for example, crested tits and willow tits apparently store food in the fall and draw on it over the winter, much as Clark's nutcrackers do. Marsh tits store superabundant food for

Figure 3.9 MARSH TIT is a small British bird (related to the North American chickadees) that caches seeds if there are more available than it needs immediately for food. Typically the marsh tit returns to draw on its caches within a few hours or at most a few days after it deposits them.

relatively short periods (see Figure 3.9). Blue tits and great tits do not store food.

Among the corvids an example of several related species that differ in their tendency to hoard has been described by Vander Wall, working with Russell P. Balda of Northern Arizona University. Clark's nutcracker lives in the same areas of the western U.S. as three other members of the crow family do, namely the piñon jay, Steller's jay and the scrub jay. The four species differ greatly in their dependence on stored piñon seeds, and concomitantly they differ in how well they are able to harvest and transport the seeds (see Figure 3.10).

Clark's nutcracker is the most dependent. By exploiting its stored seeds it is able to begin breeding as early as February, well before most other birds in the area. Clark's nutcracker is also the most specialized anatomically for harvesting and hoarding seeds. In addition to the sublingual pouch in which it carries large numbers of seeds (see Figure 3.11) it has a long, sharp bill, with which it pries open green pine cones before the cones would have opened by themselves. Moreover, Clark's nutcracker is a strong flier. It can carry heavy loads great distances to suitable storage sites.

The scrub jay, in contrast, is the least specialized

Figure 3.10 ANATOMICAL SPECIALIZATION of four birds corresponds to their varying degree of dependence on cached food. Clark's nutcracker (*a*) is the most dependent and also the most specialized: in addition to a sublingual pouch it has a long, sharp bill, with which it pries open pine cones. Moreover, it is a strong flier capable of going a great distance burdened with seeds. The piñon jay (*b*) and Steller's jay (*c*) are intermediate in both behavior and anatomy. They carry seeds in their distensible esophagus, not in a specialized pouch. The scrub jay (*d*) is the least dependent on stored seeds and least specialized in anatomy; can carry seeds only in its mouth and bill. The seed-carrying capacity of each bird is indicated by a broken colored line.

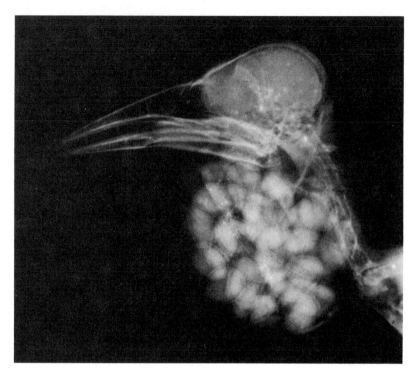

Figure 3.11 SUBLINGUAL POUCH of a Clark's nutcracker is a pocket under the tongue in which the bird carries seeds. It is thus an anatomical specialization that makes the nutcracker well suited for its life of transporting and hoarding seeds. When Vander Wall made this X-ray photograph, the pouch contained 38 seeds with a net weight of 30.6 grams, or more than an ounce.

of the four species. It lacks the anatomical adaptations of a Clark's nutcracker, and it is much less reliant on stored seeds. The seeds it does store it caches entirely within the area where it lives: the area it defends as its own territory. The piñon jay and Steller's jay are intermediate in both anatomy and behavior. Is the spatial memory of the birds of these three species less highly developed than it is in a Clark's nutcracker? The answer awaits further work in the laboratory and the field.

T he investigations of memory and food hoarding I have described are at the interface between zoological and psychological studies of animal behavior. Such research is increasing as more psychologists become convinced that it is productive to view laboratory studies of learning and memory as ways of analyzing how animals solve the problems confronting them in nature. At the same time zoologists are becoming interested in aspects of foraging behavior that clearly require a capacity for learning and memory. The joint efforts can therefore combine the zoologist's understanding of how behavior is adapted to the environment with the psychologist's conceptual tools and experimental techniques. The efforts are the beginning of exciting new developments in the study of memory.

From Bird Song to Neurogenesis

Studies of song-control centers in the canary brain reveal that new nerve cells are born in adulthood and that they can replace older cells. Such neurogenesis could hold the key to brain self-repair in humans.

. . .

Fernando Nottebohm
February, 1989

One of the most firmly established beliefs in the neurosciences has long been that all neurons, or nerve cells, in the brain of vertebrates are formed early in development, when the brain itself is growing. An adult vertebrate, it was believed, must make do with a fixed number of neurons. Hence neurons lost through disease or injury are not replaced, and learning takes place not by putting new cells into the neural circuits that control behavior but by modifying connections among a limited number of neurons.

Joseph Altman of Purdue University first challenged this belief in the early 1960's on the basis of experiments he did on rats and cats. He argued that his experimental results showed that some types of neurons continued to be formed in parts of the animals' brain even after they had reached adulthood. Altman's results, however, were not conclusive, and neurogenesis (the birth of neurons) in the adult mammalian brain is still not generally accepted.

New evidence for such neurogenesis in another class of vertebrates has now come from a most unexpected quarter: the study of how birds learn to sing. Recent experiments carried out by my colleagues and me show not only that neurons are constantly being born in the brain of birds after the birds reach maturity but also that the newborn neurons can in some cases replace older ones. Although we have not yet shown exactly what the new neurons do, we think they are used to acquire new information. Song learning by both juvenile and adult birds may therefore depend on the availability of young neurons that can be used to build novel circuits. These findings raise questions about the general stability of the neural circuits in the brain. Perhaps their most tantalizing aspect, however, is that they might eventually lead to the identification of factors that could stimulate a human brain to repair itself by replacing damaged neurons with new ones.

Taxonomists recognize some 30 different orders of birds, containing a total of about 8,500 living species. Almost half of the species are classified into the songbird suborder, Oscines, of the order Passeriformes. Songbirds tend to stand out among other types of birds because of their rich and varied song. A bird sings to announce its presence to its neighbors and to claim a breeding territory. In addition birds (generally males) sing to attract mates.

It has long been common knowledge that some

birds are able to imitate the sounds they hear. Yet before the 1950's few biologists realized this talent is routinely exercised by songbirds to develop their everyday song. W.H. Thorpe of the University of Cambridge first demonstrated this fact (and thereby established a new field of research) when he described how a European songbird, the chaffinch, learned its song. He reared isolated male chaffinches in soundproof chambers equipped with speakers. In some chambers he broadcast recorded chaffinch songs, which the young chaffinches were then able to imitate. Birds that had not been exposed to the recorded songs, however, developed abnormally simple songs. Moreover, exposing the deprived birds to the "tutor" tapes after they had reached sexual maturity did not improve their singing ability.

Thorpe concluded that birds learn to sing much as humans learn to speak, that is, by imitating models provided by their elders. He also concluded that song learning for birds such as the chaffinch is limited strictly to a "critical period" sometime before sexual maturity. Subsequent work by Peter R. Marler of Rockefeller University and Klaus Immelmann of the University of Bielefeld in West Germany showed that two other songbirds, the North American white-crowned sparrow and the Australian zebra finch, are also critical-period learners. Yet not all songbirds fall into that category. Canaries, for example, can alter their song from year to year; they are called open-ended learners.

The first sounds a newly hatched canary makes are shrill and high-pitched calls that spur its parents to feed it. Such "food begging" continues even after the juvenile has left the nest; it lasts until the bird becomes completely independent of its parents at about four weeks. Thereafter the bird produces its first rudimentary attempts at singing, which are called subsong. Subsong is soft in volume and variable in structure, and it is often done while the young bird appears to doze. Charles Darwin pointed out the similarity between subsong and the babbling of human infants; both seem to be early stages of vocal practice from which emerges the full repertoire of sounds used in communication.

As subsong becomes more structured by the end of the bird's second month, it is given a new name: plastic song. Plastic song sounds a bit like the song of adult canaries but is still quite variable. It becomes increasingly stereotyped as the bird approaches sexual maturity, at about seven or eight months. Canary fanciers have long known that the quality of a young bird's song is influenced at this stage by that of its older companions.

The final song pattern of an adult male canary, known as stable song, is sung for the duration of the bird's first breeding season. Such song can be characterized by the number of different sounds (called syllables) it contains. Although a male canary only three or four months old can already vocalize 90 percent of the syllables it will employ as an adult, it is not until sexual maturity that syllables become stereotyped. Indeed, achievement of stereotypy seems to be a difficult task for canaries, since it takes them several months of practice while in the plastic-song phase.

Yet even after such practice a canary's repertoire of syllables is not permanently learned. Every year during late summer and fall (after the breeding season) the syllable stereotypy mastered a few months earlier is lost, and the bird's song becomes as unstable as the plastic song of juvenile birds. In fact, many of the learned syllables disappear from its basic song "vocabulary" and new ones are acquired, which may then be incorporated into a stereotyped song the following winter and spring. In this way adult male canaries can develop a new song repertoire every year. Such seasonal song learning is probably under hormonal influence, since peaks of new syllable addition are preceded by drops in the level of the male sex hormone testosterone in the blood.

When Thorpe did his experiments with chaffinches, nothing was known about the parts of the avian brain that control song learning. It was not until 1976 that several anatomically distinct clusters of cells that control canary song were identified in my laboratory at Rockefeller University. Such a cluster of cells is called a nucleus. (It should not be confused with the nucleus of an individual cell, which contains the cell's genetic material.)

The largest of these nuclei, the higher vocal center (HVC), lies in the bird's forebrain. The axons —the long processes characteristic of neurons— of many of the cells in the HVC extend to another forebrain nucleus, the robustus archistriatalis (RA). Many RA neurons in turn have axons that contact hypoglossal motor neurons innervating the muscles of the syrinx, the organ that actually produces song (see Figure 4.1). Since these various brain nuclei are rather distinct, it is possible to estimate their volume accurately and relate it to a particular bird's sex and age, the hormone level in its blood and the com-

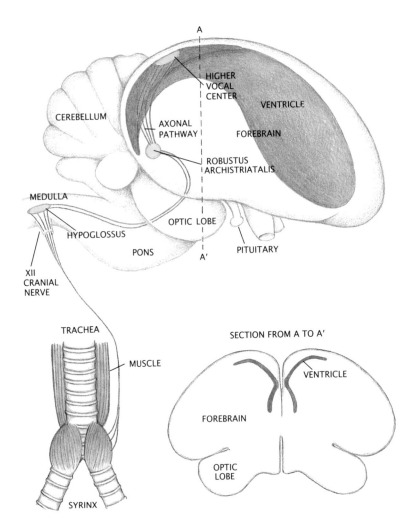

Figure 4.1 CANARY BRAIN, shown in a side view (*top*) and cross-section (*lower right*), contains several nuclei, or distinct clusters of cells, that control song learning. The largest of these nuclei is the higher vocal center (HVC). Electrochemical signals from the HVC are transmitted to other parts of the brain along axons, the long processes characteristic of neurons, or nerve cells. Many HVC neurons have axons that extend to neurons in another song-control nucleus, the robustus archistriatalis (RA). The axons of many RA neurons in turn contact motor neurons of the hypoglossal nucleus that innervate the muscles of the syrinx, the organ where sound is actually produced.

plexity of its song. Although the brain of a young canary reaches adult size sometime between 15 and 30 days after hatching (at about the time the bird becomes independent of its parents), the HVC and the RA continue to grow for several more months—almost until the bird reaches sexual maturity. It is during this period of HVC and RA growth that young birds first learn to sing (see Figure 4.2).

In 1976 Arthur P. Arnold (who was then at Rockefeller) and I discovered that the HVC and the RA were some three to four times larger in adult male canaries, which sing complex songs, than in adult female canaries, which sing simpler songs. It appeared that the amount of brain volume devoted to a particular skill was considerably greater in the sex excelling at that skill. This instance of so-called sexual dimorphism disproved another long-held view, namely that brains of vertebrates exhibited no marked anatomical differences between the sexes.

Subsequent work by Mark Gurney and Masakazu Konishi of the California Institute of Technology showed that the sexual dimorphism of the RA arises (at least in part) from differences in the number of neurons it contains and that these differences become apparent early in development—even before song learning begins. These results suggest how brain anatomy might limit learning: the greater the number (and perhaps diversity) of neurons integrated into a particular neural circuit, the greater the amount of information the circuit can handle.

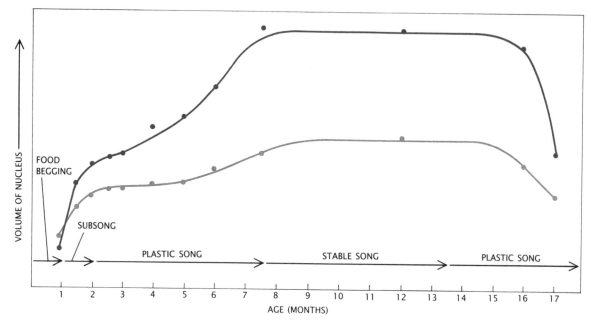

Figure 4.2 DEVELOPMENT OF SONG in male canaries is accompanied by marked increases in volumes of both HVC (*red*) and RA (*blue*). Song passes through four phases: food begging, subsong, plastic song and stable song. Each phase is more structured than the one before. After breeding, the HVC and RA shrink back to their sizes at age three months and the bird reverts to plastic song. By the next spring, they have enlarged again and stable song returns. The transitions back and forth between plastic song and stable song, with corresponding changes in nuclear volume, are repeated annually.

The relation between the size of the song-control nuclei and singing skill also holds for canaries of the same sex. Some male canaries are unusually talented singers and have developed a large repertoire of song syllables; they tend to have large HVC's and RA's. Other birds of the same breed, sex and age that have been kept under identical conditions produce simpler songs consisting of a smaller number of syllable types; they tend to have small HVC's and RA's.

Although it is tempting to infer that the anatomical differences are the basis of these marked differences in singing talent, there also are canaries that have a large HVC and nonetheless have a small syllable repertoire. An analogy between the size of the song-learning nuclei in canaries and the amount of shelf space in a library may help to explain the relation. If a library is to hold many books, it needs ample shelf space, yet the shelves of a large library need not be completely filled. (Actually, under some conditions the arrival of new "books" can expand the "shelf space." Observations by Sarah Bottjer of the University of Southern California and Arnold,

who is now at the University of California at Los Angeles, suggest that the act of song learning itself can enlarge the size of the HVC.)

Statistical analyses of correlations between the size of syllable repertoires and the size of the HVC in adult male canaries show that only 20 percent of the variability in either one can be explained as a consequence of the other. Song learning in canaries, it appears, is also influenced by factors other than HVC and RA size. Nevertheless, observations made in my laboratory and elsewhere suggest that the number of neurons in the HVC and RA (as reflected in the physical size of these nuclei) does influence the skill with which a particular canary sings.

Further evidence for the importance of the size of the HVC and RA in determining singing ability comes from the effects of testosterone. It is possible, for example, to elicit malelike song in a nonsinging adult female canary by giving it intramuscular injections of testosterone. The hormone does not merely activate existing neural circuits but doubles the volume of the HVC and RA in the female canary's brain. Similarly, in adult male canaries blood testosterone levels are very high in the spring, when their song is quite stereotyped, and low in early fall, when their song is as variable as that of juveniles (see Figure 4.3). At the same time the HVC and RA in the spring are roughly twice as large as they are in the fall.

In 1981 Timothy DeVoogd, who was then at Rockefeller, and I studied how testosterone induces growth in the RA of adult birds (there are still no data on how the hormone enlarges the HVC). The most abundant neurons in the RA send out long axons that connect with hypoglossal motor neurons innervating the syrinx. Sprouting from the main body of these neurons are many secondary branches, called dendrites, that are consistently longer in males than they are in females. Yet in the female canaries we injected with testosterone these dendrites grew and became indistinguishable from those of males. As the dendrites grew they also established more contacts, called synapses, with other neurons.

Those anatomical changes suggested that an increase in hormone level or the acquisition of a new behavior (such as singing) or both can act in adulthood to rearrange the distribution of connections between the neurons that control the behavior. Such modifications of existing neural circuits could account for the gross volume changes seen in the RA and could explain why male canaries are able to change their song in adulthood. Males of species such as the white-crowned sparrow and zebra finch, which learn their song before sexual maturity, show no gross changes in adult RA volume.

The developmental and seasonal changes in the HVC and RA volume seen in male canaries and the marked changes in HVC and RA volume induced by testosterone in adult female canaries suggest that the mastery of learned song is not a subtle process but one that requires conspicuous changes in brain circuitry. Indeed, the seasonal and hormonal changes in the size of song-control nuclei were so extraordinary for an adult vertebrate brain that my co-workers and I were compelled to ask what many neuroscientists considered an unthinkable question: Did the changes really always involve the same set of neurons—those that are present in the brain after sexual maturity?

There is an easy way to determine when a new cell is born. DNA, the substance of which genes are made, is found mainly in a cell's nucleus, and a cell that is about to divide synthesizes new DNA. Hence if one injects an animal with a radioactive form of thymidine, a DNA precursor, the thymidine is sequestered inside the nucleus of cells that are about to undergo division. When such a labeled cell divides, half of the radioactive DNA will be present in the nucleus of each of its two daughter cells, labeling them as well.

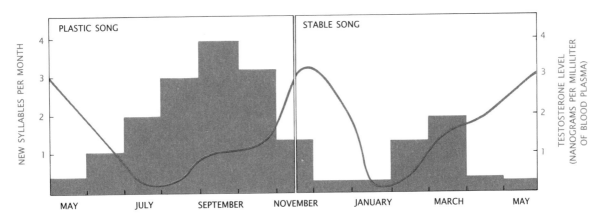

Figure 4.3 ANNUAL SONG VARIABILITY (*green***)** in adult male canaries is tied to the blood level (*orange*) of testosterone. The song of a canary can be characterized by the number of different sounds (called syllables) it contains. When a bird is in the plastic-song phase, it has not yet settled on a stereotypical song, and it often adds new syllables to its "vocabulary." The addition of new syllables is preceded by a drop in testosterone level. Conversely, few new syllables are incorporated into a canary's vocabulary when its testosterone level is high and it is in the stable-song phase.

Steven A. Goldman and I injected radioactive thymidine into adult male and female canaries every day for several days, stopped the injections and then waited for 30 days. To our surprise, when we then examined the HVC of the birds, we found that as many as 1 percent of the HVC neurons were labeled for each day the birds had received the injections. In another experiment we examined the brains of adult canaries only a day after they had received an injection of radioactive thymidine. In that case we found no labeled neurons in the HVC; we did, however, find many labeled cells in the so-called ventricular zone that overlies the HVC and forms the floor of the lateral ventricle.

These results suggested that the new HVC neurons had been born at the time of the thymidine treatment and that they originated from cells outside the HVC, in the ventricular zone. It appeared that ventricular-zone cells divided into daughter cells that migrated into the HVC, where after a period of between 20 and 30 days they turned into neurons. It so happens that neurons are born in the ventricular zone during development in birds as in all other vertebrates; neurogenesis in adulthood can therefore be viewed as no more than the retention of a developmental trait.

The neurons that develop from ventricular-zone cells look no different from other normal adult-canary neurons. Gail D. Burd and I showed that the new HVC neurons establish synaptic contacts, and John A. Paton and I showed that the new cells generate typical electrical signals when they are stimulated by other neurons. Clearly the new neurons become connected to existing neural circuits as they are added to the adult HVC.

Further studies have shown that new neurons are added constantly to the HVC of adult male and female canaries. Why then does the HVC show no growth from year to year? The obvious answer is that the new neurons replace older ones, which presumably are discarded. Not even the new neurons themselves are exempt from such replacement. My co-workers and I have found very few labeled HVC neurons eight months after the injections of radioactive thymidine, implying that most of the cells do not live much longer than that. Changes in the rates at which new neurons are born and older neurons die could contribute to the seasonal changes in volume of the song-learning nuclei I mentioned above: the number of HVC neurons drops by 38 percent at the end of the breeding season but is fully restored by the following spring.

The next challenge facing investigators is to find out just which neurons are discarded and why. Although neurogenesis and neuron replacement were discovered in a part of the adult canary brain involved in the control of song, it is by no means clear what role the new neurons play in song learning. For instance, the percentage of labeled neurons in the adult HVC per day of treatment with radioactive thymidine is at least as high in nonsinging female as it is in singing male canaries. Labeled neurons have also been found in the HVC of adult male zebra finches that were injected with radioactive thymidine after the end of the critical period, when no song learning takes place.

On the basis of these observations, it seems unlikely that the neurons added to the HVC are just components of the circuits that control the motor skills necessary for song learning. Indeed, physiological experiments suggest that the HVC not only plays an important role in song production but also may have a critical role in song recognition—a perceptual function. Because the HVC may act as a repository of perceptual memories, the addition of new HVC neurons may be necessary for birds to recognize new songs. Just as male songbirds must first acquire a perceptual memory of a model song in order to imitate it, so female birds, as well as male birds past the critical period for song learning, must acquire perceptual memories of new songs in order to recognize the sounds of mates or other birds.

Why must songbirds constantly replenish the neurons in their brain? After all, humans are thought to master motor and perceptual skills with a limited number of irreplaceable neurons. Changes in the synapses of existing neurons could well supply all the neural-circuit flexibility we need for the acquisition of new memories.

Work in my laboratory shows that synaptic changes actually do take place in some parts of the song-control system, such as the RA, where they may relate to motor learning. Yet that flexibility may not be enough for other kinds of learning. Evidence from other laboratories suggests that the kinds of input a brain cell receives can determine which of its genes are expressed, affecting the identity of a cell and the functions it performs. In some cases such genetic changes may underlie a type of irreversible learning. In other words, certain types of neurons involved in song learning may be permanently modified by the memories they hold. Hence the number of available neurons in the avian brain might limit the amount of song learning that can take place. Periodic replacement of HVC neurons may therefore be required in songbirds for them to update perceptual memories of songs.

Neurogenesis in the adult avian brain is not confined to the HVC, although that is the only part of the song-control system to exhibit the phenomenon; it is actually widespread throughout much of the avian forebrain. Interestingly, the forebrain is thought to be the part of the brain most involved with the control of complex learned behaviors.

Although neurogenesis has been reported in some adult mammals, its occurrence there seems to be much more limited and, as I indicated earlier, more controversial. Why is it so prominent in birds? It may have to do with their relatively long life span

and airborne lifestyle. A canary weighs as much as a mouse but lives 10 times longer. If a bird had to carry all the brain cells it would need to process and store the information gathered over a lifetime, its brain would have to be substantially larger and heavier.

The widespread occurrence of neurogenesis in the adult canary forebrain raises another important question. How does a new neuron find its way from its birthplace to its final position in a neural circuit? The answer may be relatively simple for neurons migrating to final sites in the HVC. Since the neurons are generated in the ventricular zone overlying the HVC, they have to traverse distances of no more than half a millimeter before they reach their final positions. Yet many of the new neurons found elsewhere in the avian forebrain are five or six milli-

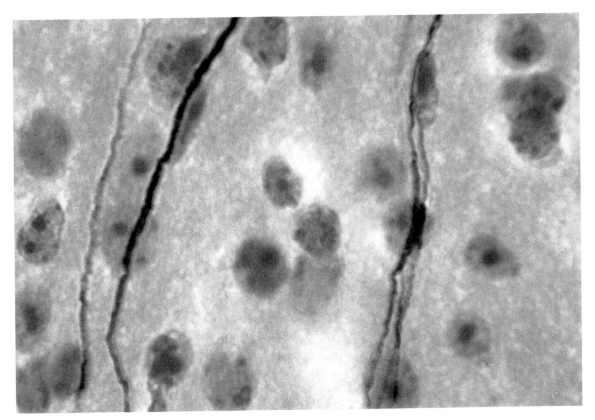

Figure 4.4 NEWBORN NEURONS adopt a characteristic elongated form as they migrate through the canary forebrain. Initially the neurons follow the long fibers (*brown*) of radial glia, a common type of cell in the avian brain. Both old and new brain cells have been made visible by staining their cell nuclei purple; two migrating cells can be seen, each on a different fiber.

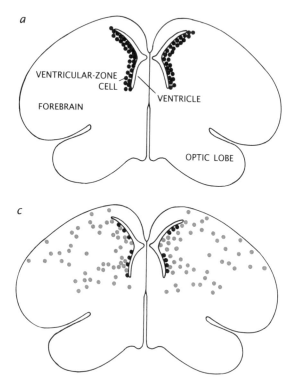

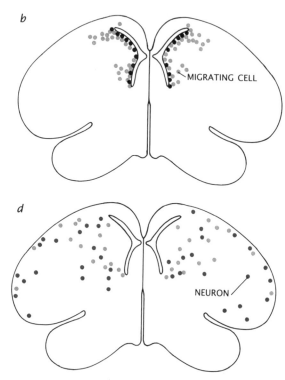

Figure 4.5 MIGRATION AND DIFFERENTIATION of neurons are mapped in this set of cross-sections of a canary brain (see Figure 4.1) one day (*a*), six days (*b*), 15 days (*c*) and 40 days (*d*) after the cells' birth. The maps are based on data gathered by Arturo Alvarez-Buylla and the author from birds injected with radioactive thymidine, a DNA precursor. The thymidine is taken up by cells that are about to undergo division and is passed on to their daughter cells, which are thereby labeled.

meters from the nearest potential birth site. What clues guide a migrating neuron as it travels through the adult brain over a distance that may be 100 times its body length?

Arturo Alvarez-Buylla and I have observed that young neurons migrating from the ventricular wall become elongated and often follow the long fibers of cells known as radial glia, which are very common in the young developing brain of vertebrates as well as in the adult avian forebrain (see Figure 4.4). Although the body of a radial glial cell is lodged in the ventricular wall, its fiber extends into the bulk of the forebrain's gray matter. After several days of migration a young neuron detaches itself from the fiber. We think this occurs when a young neuron approaches the general area where it is meant to assume its adult shape and become part of an existing circuit. Only a third of the migrating cells actually undergo the transformation into a functional, fully differentiated neuron; the rest disappear. A cell's migratory phase lasts for only a few weeks, and it may be that migrating neurons that lose their way during their journey or fail to find a position in a neural circuit in that time simply perish (see Figure 4.5).

Our studies have shown that—contrary to a long-standing doctrine of neurobiology—the brain cells of some adult vertebrates are indeed replaceable. Neurons can be born in adulthood, can travel through the adult vertebrate brain and can take their place in the neural circuits that underlie learning. If the same process could take place in the human brain, it would be invaluable for repairing neural circuits damaged as a result of disease or injury.

Yet there is no evidence that neurogenesis occurs

in man or other primates. Perhaps the reason is that humans thrive in the recollection of events long past, and neuronal replacement would disrupt such memories. Nevertheless, it may be possible to induce neurogenesis in adult brains where it does not normally take place. After all, the same genes that orchestrate neurogenesis in the young, developing brain should still be present in the brain cells of an adult. The challenge is to identify the genes and to activate them.

The Hearing of the Barn Owl

The bird exploits differences between the sound in its left and right ears to find mice in the dark. It can localize sounds more accurately than any other species that has been tested.

• • •

Eric I. Knudsen
December, 1981

For the barn owl life depends on hearing. A nocturnal hunter, the bird must be able to find field mice solely by the rustling and squeaking sounds they make as they traverse runways in snow or grass. Like predators that hunt on the ground, the barn owl must be able to locate its prey quickly and precisely in the horizontal plane. Since the bird hunts from the air, it must also be able to determine its angle of elevation above the animal it is hunting (see Figure 5.1). The owl has solved this problem very successfully: it can locate sounds in azimuth (the horizontal dimension) and elevation (the vertical dimension) better than any other animal whose hearing has been tested.

What accounts for this acuity? The answer lies in the owl's ability to utilize subtle differences between the sound in its left ear and that in its right. The ears are generally at slightly different distances from the source of a sound, so that sound waves reach them at slightly different times. The barn owl is particularly sensitive to these minute differences, exploiting them to determine the azimuth of the sound. In addition the sound is perceived as being somewhat louder by the ear that is closer to the source, and this difference offers further clues to horizontal location. For the barn owl the difference

in loudness also helps to specify elevation because of an unusual asymmetry in the owl's ears. The right ear and its opening are directed slightly upward; the left ear and its opening are directed downward. For this reason the right ear is more sensitive to sounds from above and the left ear to sounds from below.

The differences in timing and loudness provide enough information for the bird to accurately locate sounds both horizontally and vertically. To be of service to the owl, however, the information must be organized and interpreted. Much of the processing is accomplished in brain centers near the beginning of the auditory pathway. From these centers nerve impulses travel to a network of neurons in the midbrain that are arranged in the form of a map of

Figure 5.1 STRIKE OF THE BARN OWL in total darkness is shown in this sequence of exposures made in the laboratory with infrared radiation, to which the eyes of the owl are not sensitive. Unlike ground-living predators, the bird must determine its elevation above the prey as well as the direction in the horizontal plane. The owl can locate mice solely by sound. Moreover, just before striking it aligns its talons with the long axis of the mouse's body, as is shown in the final exposure. This action shows that the bird can infer the direction of the prey's motion from sound.

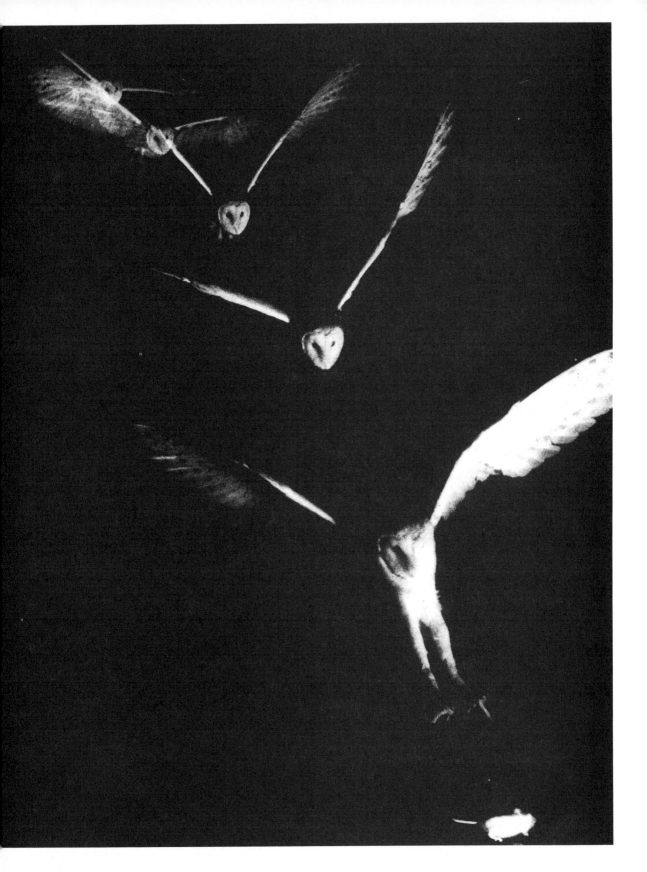

space. Each neuron in this network is excited only by sounds from one small region of space. From this structure impulses are relayed to the higher brain centers. The selection of sensory cues and their transformation into a map of space is what enables the barn owl to locate its prey in total darkness with deadly accuracy.

The barn owl has a wide range, both as a species and as an individual hunter. Barn owls are found throughout the tropical and temperate areas of the world. Many live close to human settlements, often nesting in barns or in belfries; they also nest in hollow trees and in holes in earth banks or rocks. Like most other owls they remain paired for long periods, sometimes for life, returning to breed in the same place year after year. The birds hunt in open areas, and they cover more ground than any other nocturnal bird. Studies of the bird's pellets (small objects coughed up by the bird that contain the indigestible remnants of prey) have shown that more than 95 percent of its prey are small mammals, mainly field mice; the rest are amphibians and other birds.

The nine species of barn owls are different enough from other owls to form their own family: the Tytonidae. The common barn owl, *Tyto alba*, is the most numerous species. It stands between 12 and 18 inches high and has a white face, a buff-colored back and a buff-on-white breast; its lower parts are mostly white with dark flecks. Each of the bird's middle toes has a small comb with which it dresses its feathers.

The most striking anatomical feature of the barn owl, and the one that plays the most important role in its location of prey, is the face (see Figure 5.2). The skull is relatively narrow and small and the face is large and round, made up primarily of layers of stiff, dense feathers arrayed in tightly packed rows. The feathered structure, called the facial ruff, forms a surface that is a very efficient reflector of high-frequency sounds.

Two troughs run through the ruff from the forehead to the lower jaw, each about two centimeters wide and nine centimeters long. The troughs are similar in shape to the fleshy external pinna of the human ear, and they serve the same purpose: to collect high-frequency sounds from a large volume of space and funnel them into the ear canals. The troughs join below the beak but are separated above it by a thick ridge of feathers. The ear openings themselves are hidden under the preaural flaps: two

flaps of skin that project to the side next to the eyes. The entire elaborate facial structure is hidden under a layer of particularly fine feathers that are acoustically transparent. The acoustic properties of the facial ruff are closely associated with the bird's method of locating the source of the sound.

In order to survive the barn owl must be able to locate prey with sound alone. Field mice are difficult to see even in broad daylight because their coloring blends with that of their surroundings; in addition they tend to travel through tunnels in grass or snow. By night, when the mice forage, they are essentially invisible even to the keen eyes of the owl. Hunting from the air makes the task even more difficult, since the owl must determine the angle of elevation above the prey. A determination of azimuth alone would leave an entire line of possible target sites along the ground below.

The barn owl locates sounds in two spatial dimensions with great accuracy. Roger S. Payne, and later Masakazu Konishi and I, demonstrated that the bird is capable of locating the source of a sound within a range of one to two degrees in both azimuth and elevation; one degree is about the width of a little finger at arm's length. Surprisingly, until the barn owl was tested, man was the species with the greatest known ability to locate the source of a sound: human beings are about as accurate as the owl in azimuth but are three times worse in elevation. Monkeys and cats, other species with excellent hearing, are about four times worse than owls in locating sounds in the horizontal dimension, the only one in which they have been tested.

The sensitivity of the barn owl's hearing is shown both by its capacity to locate distant sounds and by its ability to orient its talons for the final strike. When the owl swoops down on a mouse, even in a completely dark experimental chamber, it quickly aligns its talons with the body axis of the mouse. It was Payne who first suggested that this behavior is not accidental. When the mouse turns and runs in a different direction, the owl realigns its talons accordingly. This behavior clearly increases the probability of a successful strike; it also implies that the owl not only identifies the location of the sound source with extreme accuracy but also detects subtle changes in the origin of the sound from which it infers the direction of movement of the prey.

Several kinds of experiments have helped to elucidate how the barn owl accomplishes these diffi-

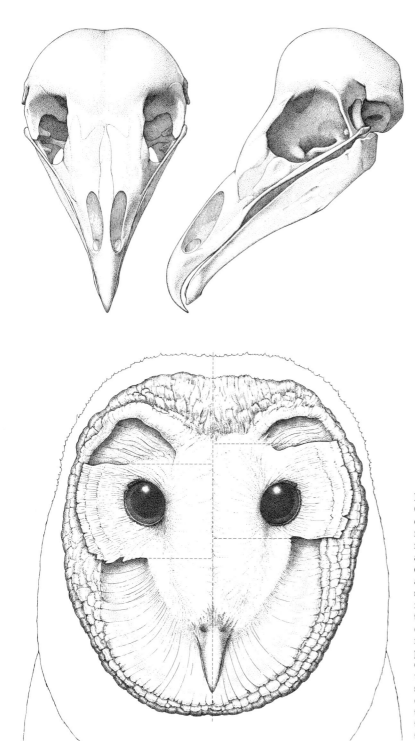

Figure 5.2 FACE OF THE BARN OWL is formed by the facial ruff extending from the relatively narrow skull. The external ears are feather troughs that collect sounds and funnel them into ear-canal openings hidden under the preaural flaps. The left ear is more sensitive to low-frequency sounds from the left and the right ear to those from the right. At high frequencies, however, the right ear is more sensitive upward, because the preaural flap and opening are lower on the right and the trough is tilted up. The left ear is more sensitive to sounds coming from below. Differences in perceived loudness can therefore yield clues to the elevation of a source of sound as well as to its horizontal direction.

cult tasks. Experiments with birds in free flight have measured how accurately the owl flies toward and strikes at an invisible sound source. Head-orientation experiments, where the bird was perched on a testing stand, have helped to measure the precision with which it aligns its head with an incoming sound (see Figure 5.3). Experiments where the brain of an anesthetized owl was probed with a microelectrode while the bird was exposed to various sounds have shown how the sensory information is organized and interpreted by the central nervous system.

Konishi and I have done a series of head-orientation experiments. This experimental system has several advantages over tests conducted with birds in free flight. In free-flight tests flight errors may be confused with sound-location errors. In addition the angle of the sound source in relation to the bird's head at the moment the bird decides to strike cannot be determined. Free-flight trials are also complicated and time-consuming to conduct. In contrast, head-orientation experiments are relatively simple to conduct, and they allow the relation between the head and the sound source to be measured.

These experiments take advantage of a natural response of the owl when it is hunting. On hearing a noise the bird turns its head in a rapid flick that brings the source of the sound directly in front of it. This movement brings the sounds into the region where the bird's hearing is keenest. The eyes of the barn owl are immobile, and so the movement also enables the bird to see a target with maximum acuity. Konishi and I monitored this behavior by mounting a lightweight "search" coil on top of the

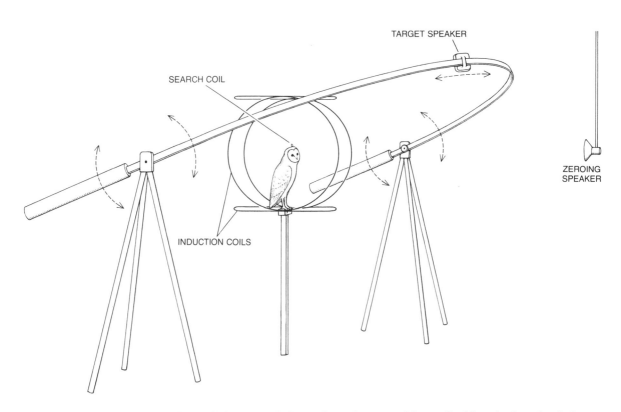

Figure 5.3 HEAD-ORIENTATION TESTS measured the accuracy of the owl's hearing by taking advantage of an owl's natural tendency to face a sound source. The perching owl had a "search" coil mounted on its head, which was centered within magnetic fields from stationary induction coils. Any head movement caused a measurable change in search coil current. The owl's attention was first directed to a sound from a fixed "zeroing" speaker before a new sound from the movable "target" speaker made the bird turn its head suddenly. A computer controlled the location of the target speaker and recorded the head movements, allowing calculation of the owl's ability to orient accurately.

owl's head. Magnetic fields generated by other coils were centered so that when the bird perched normally, the search coil was at the intersection of the horizontal field and the vertical one. The electric current induced in the search coil varied with its orientation to these fields. By evaluating the magnitude of two distinguishable signals one could measure the horizontal and vertical components of the orientation of the owl's head (see Figure 5.4).

The tests were done in a totally dark chamber lined with materials that eliminate echoes. Sounds were generated by a stationary speaker (the "zeroing" speaker) and a movable speaker (the "target" speaker). The owl first turned to face a sound from the zeroing speaker, placed in front of its perch; a sound delivered by the target speaker then caused the bird to turn its head in the characteristic flicking movement. A computer controlled the location of

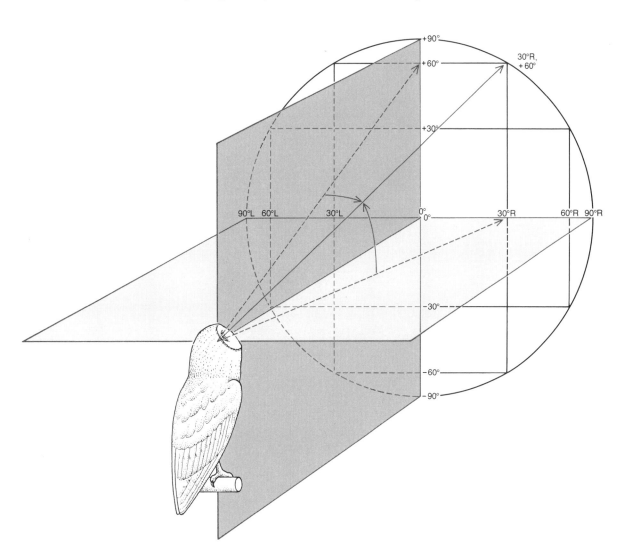

Figure 5.4 DOUBLE-POLE COORDINATE SYSTEM serves to map the positions of the owl's head in head-orientation tests. Each direction is specified by two measurements: the angle from the horizontal and that from the vertical through which the owl must turn its head to face that way. Facing forward, as the bird normally perches, the orientation is 0 degrees (horizontal) and 0 degrees (vertical). The map represents the 180 degrees of space in front of the bird. The other maps accompanying this chapter have the same coordinate system.

the target speaker and recorded its location and the alignment of the owl's head.

The head-orientation trials have yielded much information about how the owl determines the origin of a sound. One of the important features of the process is that it can achieve maximum accuracy even with sounds that end before the head movement begins. This indicates that the owl's auditory system determines the azimuth and elevation of a sound without head movement and then utilizes the information to direct the head-orientation response. The head-movement tests also show that the owl's accuracy deteriorates with increases in the angle between the source of the sound and the orientation of the bird's head.

Our experiments and those of others have shown that the barn owl's ability to locate the origin of a sound is dependent on the presence of high frequencies in the sound. Although the owl's hearing is sensitive to a broad range of frequencies, from 100 hertz (cycles per second) to 12,000 hertz, it can locate accurately only sounds with frequencies between 3,000 and 9,000 hertz. In addition experiments in which one of the bird's ears is plugged show that both ears are necessary for the accurate locating of targets. If one ear is plugged, the owl makes large errors in the direction of that ear.

With these characteristics in mind we proceeded to investigate the exact information the barn owl selects from natural sounds. To locate the source of a sound the owl must determine the direction of propagation of the sound waves based on information from detectors at two points, namely its ears. The most useful spatial information is gained by comparing information from these two sources, since the differences between them depend not on the absolute sound level but only on the orientation of the ears in the sound field.

One valuable cue of this kind is the difference in the time of arrival of the sound in the two ears. When the sound comes directly from the side, the difference is at its maximum; when the sound is directly in front of the bird, there is no difference in the arrival time at the two ears. Between these limits the time difference varies with the angle of the sound in the horizontal plane. The time delay can therefore yield information about the azimuth of the sound. This is not sufficient for the bird to locate the sound exactly, because sounds from several directions can give rise to the same difference in time. In three-dimensional space these directions form a cone around the axis between the owl's ears (see Figure 5.5).

This is an example of the inadequacy of any one cue to provide information about both the horizontal and the vertical angle of a source of sound. To specify a location in both dimensions two independent cues are needed. In the owl's case the additional information is provided by differences in the directional sensitivity of the ears.

Directional sensitivity is provided by the facial ruff. Like a hand cupped behind the ear, the troughs of the ruff amplify the sound and make the ear more sensitive to sounds from certain directions. The amount of amplification and directional sensitivity imparted by the feathers of the facial ruff varies dramatically with the frequency of the sound (see Figure 5.6). This is owing to one of the properties of the sound waves themselves. When sound waves encounter an object, they can bend around it or be reflected back from it. Which of these happens depends on the wavelength of the sound and the size of the object. If the wavelength is long compared with the object, the waves tend to propagate around the object; if the wavelength is short, the waves tend to be reflected back in the direction from which they came.

As a result of this phenomenon frequencies of less than 3,000 hertz are not efficiently reflected by the ruff. Because the funneling action of the ear depends on its capacity to reflect sound its directional sensitivity at low frequencies is relatively poor. At 3,000 hertz for example, the left ear is only slightly more sensitive to sounds coming from an area between 20 and 40 degrees to the left than it is to sounds coming from other directions. The right ear has a similar degree of sensitivity to the right. Since the sensitivity of each ear at low frequencies changes only gradually with direction, the comparison of sound intensities at low frequencies can provide only a coarse spatial cue. Moreover, this difference yields no clue to the elevation of the sound.

With higher-frequency sound waves the situation is quite different. Each ear is much more sensitive to the direction of the sound; a small change in sound direction gives rise to a large change in perceived intensity. In addition, instead of being more sensitive to the right or to the left, the right ear is more sensitive above the horizontal plane and the left ear is more sensitive below it. This sensitivity is the result of an unusual anatomical asymmetry. The ruff on the left is directed slightly downward, and

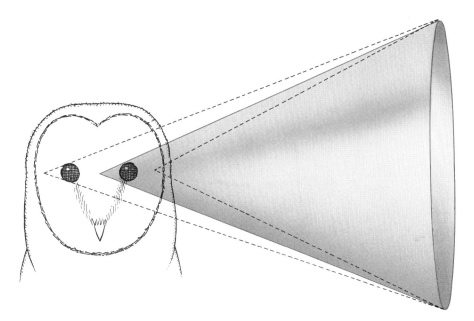

Figure 5.5 CONE OF CONFUSION (*color*) is formed by the directions among which the barn owl cannot distinguish on the basis of time delay alone. Time delay can provide information about the horizontal angle of a sound source. Since the left and right ears are generally at slightly different distances from the source of a sound, sound waves reach them at slightly different times. The greater the angle of a sound source from the bird's frontal plane, the greater the time delay. There are many directions, however, that give rise to the same path lengths (*broken lines*) and hence to the same time delay; other cues are required for the bird to tell them apart. In three-dimensional space these directions form a cone whose peak is between the two ears.

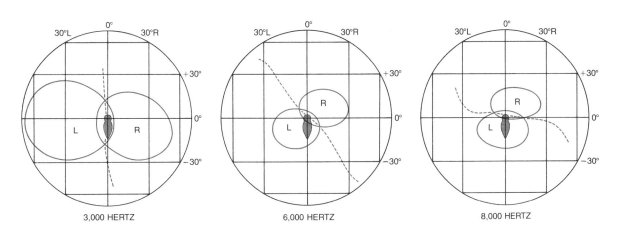

Figure 5.6 DIRECTIONAL SENSITIVITY of barn owl ears changes with the frequency of sound waves. With low frequency sounds, each ear is most sensitive to the side. Because high frequencies are more efficiently reflected by the facial ruff, they are more affected by the upward (right ear) or downward (left) tilts. As frequency rises, the areas of greatest sensitivity become increasingly vertically aligned becoming almost directly above and below the bird at 8,000 hertz (cycles per second). Because natural sounds usually have many frequencies, the owl can exploit differences in loudness to identify both the horizontal and vertical directions of a source.

the ear opening and the preaural flap are higher in the ruff on the left side. On the right side the reverse is the case. Accordingly as the source of the sound moves up the high-frequency components of a natural sound become louder in the right ear and softer in the left. As the source of the sound moves down the sounds become louder in the left ear. Since at high frequencies the perceived loudness changes rapidly with elevation, this cue offers information that is very precise.

Although that information is valuable to the barn owl, it is also complex. The magnitude of the difference in intensity varies according to the frequency of the sound, because of the greater capacity of the ruff to reflect sounds at higher frequencies. The direction indicated also varies with frequency, since horizontal location is given by low frequencies and vertical location by high ones. The owl's auditory system must therefore compare the intensities detected at each ear for each frequency. A comparison of intensities made frequency by frequency is called an interaural spectrum.

Since low-frequency sounds yield clues to azimuth and those of high frequency yield clues to elevation, the interaural spectrum could by itself provide enough information for the owl to locate prey. Much evidence supports the hypothesis that owls use this spectrum. The strike accuracy of the bird increases sharply as the bandwidth (the number of frequencies contained in a sound) is increased. From differences in the intensity of a single tone (a sound of only a single frequency) the owl can determine direction in only one dimension; as the spectrum broadens more intensity differences are available, and their values indicate the angle of the source in more than one plane. Experiments in which one of the owl's ears is plugged show more directly that the owl compares intensities. A trained owl with its right ear plugged strikes to the left and short of the target; one with its left ear plugged strikes to the right and beyond the target. That the errors have components of both elevation and azimuth implies that the owl gains both types of information from comparisons of intensity.

Further confirmation has been obtained by removing the owl's facial ruff. When the ruff is removed, the owl is able to locate the azimuth of a sound quite well but cannot identify its vertical location: the bird consistently orients to a point on the horizontal, regardless of the elevation of the target. This accords with our hypothesis: the sound-reflecting properties of the ruff underlie directional sensitivity at high frequencies, which enables the owl to identify the vertical angle of the source. Removal of the ruff eliminates the ability to discriminate among elevations. Some disparity in intensities is retained because the ear openings and preaural flaps are placed asymmetrically in the fold of skin supporting the feathers of the ruff, but this is not enough to make it possible to identify the elevation of the sound.

That barn owls rely on the interaural spectrum to locate sounds in both azimuth and elevation has thus been amply confirmed by the results gathered from the head-orientation tests. It is clear from other findings, however, that the bird also makes some use of timing differences in locating its invisible prey. The timing delay is manifested in two aspects of the binaural signal. First, the sound begins and ends sooner in the ear close to the source; the timing of major discontinuities in intensity in the sound is also slightly different in each ear (see Figure 5.7). These differences are known collectively as transient disparity. Second, throughout the duration of the sound the sound waves reaching the far ear will be slightly delayed. With a single frequency this difference in the timing of the waveforms is known as phase delay; with more complex natural sounds, made up of many frequencies, it is called ongoing time disparity.

In nature the ongoing and the transient disparities are of about equal magnitude; they vary with changes in the azimuth of the source of sound. They do, however, have different advantages for locating sounds. The ongoing time disparity can be measured repeatedly while the sound lasts. Transient disparity, on the other hand, can be monitored only intermittently, but it is less likely than the ongoing disparity to be confused by echoes.

Human beings rely on both transient and ongoing disparity to determine the source of a sound. Owls appear to rely on only the ongoing difference. Like other kinds of spatial information, the ongoing disparity has a major ambiguity. It may be best understood in the case of a tone coming directly from the side. The signals detected by the owl's ears are sinusoidal (regular) waves; because of the different distances to the ears the waves will be slightly out of phase with each other. The magnitude of the phase delay so created will depend both on the frequency of the tone and on the distance between the ears. As the frequency of the wave or the distance between the ears increases, the wave passes through more of

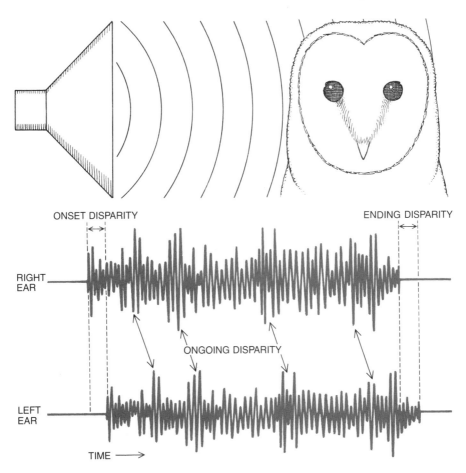

ONSET DISPARITY

ENDING DISPARITY

RIGHT
EAR

ONGOING DISPARITY

LEFT
EAR

TIME ⟶

Figure 5.7 SOUND TRACE shows two kinds of differences in the timing of the sound between the barn owl's left and right ears. In this schematization the height of the trace indicates the pressure caused by the sound waves in each ear. When a sound comes from the right, the waves begin and end earlier in the right ear; major changes in intensity also occur slightly earlier there. These differences are collectively known as transient disparity. In addition, throughout the duration of the sound the waveforms are slightly advanced in the right ear; this is called ongoing disparity. Man relies on both differences to find the source of a sound. The barn owl relies heavily on ongoing disparity, but there is no evidence that the bird exploits the transient difference.

its cycle as it travels around the head to reach the far ear; hence the phase delay is greater.

When the frequency of the tone is so high that the wave passes through exactly half of its cycle before it reaches the far ear, the phase delay corresponds to half of the wavelength. Such a delay could be caused by a sound coming directly from the owl's left or directly from its right, since the difference in path lengths is the same for these two directions. It is therefore impossible for the bird's auditory system to determine the direction of the sound on the basis of ongoing disparity alone. At higher frequencies the situation is worse still. When the wavelength is equal to the distance between the ears, for example, there is no phase delay, since the wave travels through its entire cycle while passing around the head. This relation could correspond to a sound coming from the right, from the left or from directly ahead; on the basis of phase delay alone the bird has no way of determining which is the case.

The wavelengths at which such ambiguities arise depend on the distance between the ears. The barn owl's ears are about five centimeters apart; phase ambiguity will therefore arise at a wavelength of 10 centimeters or less, which corresponds to frequencies of 3,000 hertz or more. Since the barn owl has no difficulty determining the azimuth of even high-frequency tones, Konishi and I assumed that at high frequencies the bird must rely on some source of information other than the ongoing time disparity. A likely candidate for the additional cue was transient disparity, because it is not affected by changes in frequency and because other species, including man, are known to depend on it at high frequencies.

How wrong this conclusion was has been demonstrated conclusively by Andrew Moiseff and Konishi in a further head-orientation experiment with the barn owl. They presented sound directly and independently to both ears by means of small speakers implanted in the owl's ear canals. This technique enabled them to eliminate transient disparities and differences in intensity between the ears as they varied the ongoing time disparity. The delay between the waveforms of the sound in the two ears could be adjusted in steps as small as one microsecond. In response to ongoing disparities of as little as 10 microseconds or as much as 80 microseconds the owl made quick horizontal turns of the head that corresponded approximately to the angle implied in the ongoing difference. This response suggests, in the absence of other interaural cues, that the owl continues to make use of the phase delay, or ongoing disparity, even at high frequencies.

The finding was startling because it implied that the barn owl has some means of overcoming the phase ambiguity. Furthermore, for the owl's auditory system to sense such small differences in the timing of the waveforms in the two ears it must receive specific information about acoustic signals occurring at 7,000 hertz, or once every 143 microseconds. This is remarkable, because the nerve impulses that convey this information from the cochlea to the brain last for more than 1,000 microseconds.

Ongoing time disparity is very useful to the barn owl; by itself, however, the cue does not have the precision the owl needs in order to hunt. With speakers implanted in the ear canals the owl responded to a given ongoing time disparity with turns that varied in magnitude by up to 15 degrees in either direction. In contrast, the largest standard deviation of error for owls responding to external targets is only 2.5 degrees in either direction, and the error is usually less than 1.5 degrees. Other cues

must therefore be combined with the ongoing disparity. It is known that differences in intensity help to specify location, but we also tried to discover whether the bird employs transient disparities.

Our original hypothesis, that the owl relies on transient disparity to compensate for phase ambiguity at high frequencies, has clearly been invalidated. The owl's performance in determining the azimuth of a tone suggests that transient disparities are not relied on at all. In head-orientation trials the owl was presented with tones of 7,000 and 8,000 hertz. The speaker was sometimes at 30 degrees to the right and sometimes at 30 degrees to the left. In these situations a curious kind of behavior was observed: with the target at 30 degrees to the right the owl sometimes turned 30 degrees to the left, and vice versa.

This confusion is clearly a manifestation of phase ambiguity. The cycle period of a 7,000-hertz tone is 143 microseconds and that of an 8,000-hertz tone is 125 microseconds. With a source of sound at 30 degrees to the right or the left these tones pass through about half a cycle in traveling over the difference between the path lengths to the ears. Therefore a tone coming from 30 degrees to the right yields the same phase delay as one coming from 30 degrees to the left. The ambiguity is present only in the ongoing disparity; if the owl had been relying on transient disparity, it would immediately have picked out the correct location.

This result confirms that the owl continues to rely on phase-delay information, even at high frequencies, in spite of its ambiguity, and apparently does not rely on transient disparity. Since the wavelength of low-frequency sounds is considerably greater than the distance between the ears, phase ambiguity does not exist at low frequencies. Moreover, the presence of a number of frequencies in natural sounds helps the owl to resolve phase ambiguities. When many different pairs of directions arise from different frequencies, the owl selects the one direction in space that is consistent with one member of each pair.

Head-orientation trials clearly show that the barn owl relies on two kinds of information to determine the origin of a sound: the interaural spectrum and the ongoing time difference. The latter yields clues to the azimuth of the sound source. The differences in intensity between the ears yield clues to both the azimuth and the elevation. To learn how the auditory system transforms these cues into a neural image of sound location calls for a different experimental approach.

The technique I resorted to, first with Konishi and later in independent experiments, is that of inserting a microelectrode into the brain of an anesthetized owl and searching for sites of neuronal activity while sounds are presented to the bird's ears. Konishi and I began the experiments with the same apparatus we had used in the head-orientation trials. After the owl was anesthetized its head was held rigidly in a special stereotaxic frame. A microelectrode was lowered into the brain until nerve impulses from a single neuron could be recorded. By moving the target speaker around the owl it was possible to map the regions of space to which the neuron responded.

Since the source of a sound in space is determined only after considerable neural processing, we began our study by exploring structures fairly far along the auditory pathway: in the midbrain and forebrain. The main auditory center in the midbrain of birds is called the nucleus mesencephalicus lateralis pars dorsalis (MLD). (It corresponds to the structure in the brain of mammals called the inferior colliculus.) Nerve impulses reaching this center have already been processed in one or more nuclei farther down the auditory pathway (see Figure 5.8). Farther up the pathway an area designated Field L is the primary receiving center in the forebrain for auditory impulses. (This structure corresponds to the auditory cortex in mammals; birds have no exact analogue to the auditory cortex.)

In both the MLD and Field L the large majority of neurons do not respond precisely to spatial cues. Some of the neurons show their highest level of activity in response to sounds from a certain region of space, but the borders of the region are not sharp, and they vary greatly with the intensity of a sound. Other neurons are even less specific in their response, being excited by sounds from virtually all directions. These neurons probably contribute not to specifying location but to the detection or identification of sounds.

Two types of neuron found in the MLD and Field L, however, are highly sensitive to sound location. The first of them, called the complex-field neuron, is found in large numbers in the MLD, usually in clusters scattered among other kinds of neurons. The activity of complex-field neurons is stimulated by sounds coming from several separate regions of space, called excitatory fields. The neurons are much less excited, or are even inhibited, by sounds rising in the regions of space between the excitatory fields (see Figure 5.9).

When the location of the centers of the excitatory fields in space are calculated for an individual complex-field neuron, an interesting correspondence is observed. The excitatory fields are the same as the regions of space the owl confuses under conditions of phase ambiguity. The multiple excitatory fields represent the regions from which sounds reaching the ears will generate phase delays of equivalent magnitude. Each complex-field neuron therefore seems to be sensitive to a particular phase delay at a particular frequency. The presence of several excitatory fields for each neuron would appear to be the physical correlate of spatial confusion due to phase ambiguity.

The second type of neuron, called the limited-field neuron, which is found in both the MLD and Field L, responds in an even more specific way. Limited-field cells are excited only by sounds coming from a single region of space. The regions to which limited-field neurons respond are typically elliptical. Their size varies in azimuth from seven degrees to 42 degrees and in elevation from 23 degrees to an entire band in front of the bird. Unlike other neurons in the auditory pathway, the limited-field neurons are extremely selective, responding only to changes in location; large changes in sound intensity cause little if any alteration in the sharp borders of the receptive region.

Some of the sharpness of the borders of the limited-field-neuron receptive regions is due to the fact that sounds coming from outside the excitatory region inhibit the cell's response. The inhibitory effect becomes stronger as the location of the sound approaches the border of the excitatory region, at which point the inhibiting effect changes to one of excitation. Although the spatial regions that give rise to an excitatory response are fairly large, within them there is a smaller region of the best response. These "best regions" vary in size from 2.5 to 15 degrees in azimuth and from five to 45 degrees in elevation.

Limited-field units are found in both the MLD and Field L. Their distribution in the two centers is, however, fundamentally different. In Field L limited-field neurons make up only about 15 percent of the total population of nerve cells, and they are often found scattered among other types. In the MLD, on the other hand, these cells are concen-

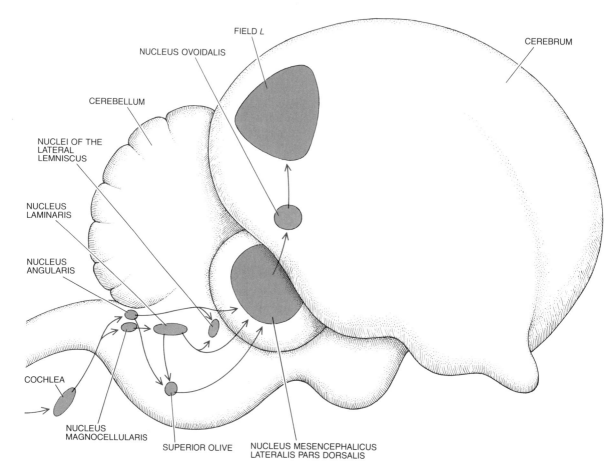

FIELD L

NUCLEUS OVOIDALIS

CEREBRUM

CEREBELLUM

NUCLEI OF THE
LATERAL
LEMNISCUS

NUCLEUS
LAMINARIS

NUCLEUS
ANGULARIS

COCHLEA

NUCLEUS
MAGNOCELLULARIS

SUPERIOR OLIVE

NUCLEUS MESENCEPHALICUS
LATERALIS PARS DORSALIS

Figure 5.8 AUDITORY PATHWAY of the barn owl leads from the cochlea to Field L. Along this route data about the timing and frequency of sound are converted into information about the location of the sound source. Much processing takes place in various centers of the lower brain. By the time the impulses have reached the mesencephalicus lateralis pars dorsalis (MLD) they are directed into a network of neurons that respond to sounds from specific areas; the distribution of those areas forms a two-dimensional map of the space in front of the bird. Information about a sound's location then passes to Field L, corresponding to the auditory cortex of mammals.

trated on the side and front margins of the nucleus, interspersed only with a few neurons of the complex-field type. After we had intensively explored and mapped the MLD it became apparent to us that the arrangement of limited-field neurons in that nucleus constitutes a map of two-dimensional space, with the distribution of the receptive areas of the neurons following the contours of space.

The map is distorted, however: the area in front of the bird, the region of maximum auditory acuity, is represented disproportionately (see Figure 5.10). Sound azimuths are arrayed in the horizontal plane.

On the right side of the structure representing the map are neurons that are excited by sounds originating between 15 degrees to the right and 60 degrees to the left. On the opposite side are neurons stimulated by sounds between 15 degrees to the left and 60 degrees to the right. This arrangement means that the 30 degrees of space in front of the bird is represented on both sides of the map and therefore by two sets of neurons. In addition the map is arranged so that on both sides the population of neurons that represent the 30 degrees in front is disproportionately large. As a result of the

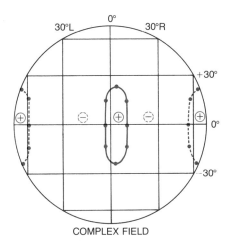
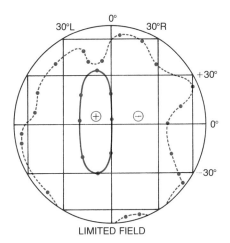

COMPLEX FIELD   LIMITED FIELD

Figure 5.9 TWO TYPES OF NEURONS in the barn owl brain are sensitive to sounds from specific directions. Shown here are maps of the receptive fields of the two cell types. Receptive fields are the regions of space within which sounds will produce an excitatory response in the neuron. Complex-field neurons, found in the MLD, have several such areas (*left*). The receptive areas correspond to the directions that give rise to identical ongoing disparities in the timing of sound waves. Between those regions sounds produce an inhibitory response. Limited-field cells, found in both the MLD and Field *L*, have a single receptive area (*right*) and strongly inhibitory zones in between. Each limited-field cell responds to a specific difference between the ears in sound timing and intensity.

double representation and the large number of neurons on each side the 30 degrees in front is analyzed with great precision, a fact that may explain the owl's particular accuracy in locating sounds in this area.

Sound elevations are arrayed transversely on the map. The "best regions" of the limited-field neurons range from 40 degrees upward to 80 degrees downward. The upper fields are at the top of the curved surface of the map and the lower fields are at the bottom.

How does the nervous system of the barn owl construct this remarkable map? Substantial understanding of the process has come from experiments by Moiseff and Konishi. In these tests speakers were placed in the ear canals of anesthetized owls. When sound was delivered separately to the two ears, the timing and intensity differences required to elicit a response in each neuron could be observed. These values were then correlated with the areas of space known to excite each set of neurons. The results of the investigations show that the limited-field units in the map are quite sensitive to ongoing differences in timing. This observation corresponds nicely to the bird's dependence on ongoing disparities demonstrated in head-orientation experiments.

Limited-field neurons respond only to an ex-

tremely narrow band of ongoing time delays: the size of the band ranges from 40 to 100 microseconds. Even within this minute range one particular delay always elicited the greatest response. Changing the ongoing difference by as little as 10 microseconds could change the strength of the neuron's response by as much as 75 percent. This degree of sensitivity complements (and helps to explain) the owl's precision in aerial hunting.

The ongoing disparity that gives rise to the greatest response also corresponds to the region of space to which the cell responds. Neurons that respond to sounds coming from the front are maximally excited by small ongoing disparities. Neurons that respond to sounds at greater angles require larger disparities. These results confirm at the level of individual cells the conclusion drawn from behavioral tests: that the owl relies heavily on the ongoing difference in timing between the ears to determine location in the horizontal plane.

Such experiments have also helped to confirm and explain the owl's reliance on differences in sound intensity. By varying the relative intensity of sounds presented independently to the ears of an anesthetized owl it was found that for each limited-field neuron there is one difference in intensity that

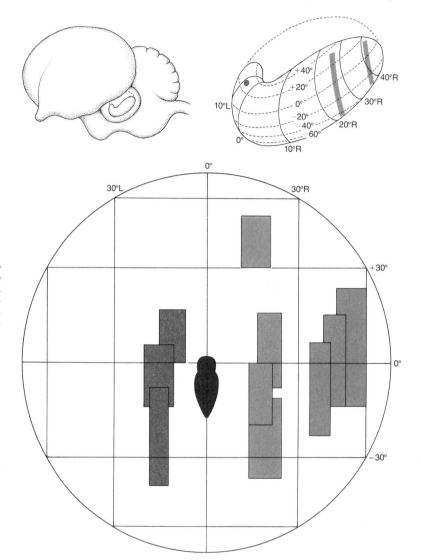

Figure 5.10 TWO-DIMENSIONAL MAP of space in the barn owl's MLD is made up of limited-field neurons that respond independently to sounds from a specific region of space. Shown above is the region on the brain's left side that responds to sounds spanning from 15 degrees on the left to 60 degrees on the right, and from 40 degrees upward to 80 degrees downward. Electrode traces down through the structure reveal a layered set of spatial regions (three are shown). A corresponding structure on the right side of the owl's brain maps sounds originating from the left. The region from 15 degrees on the left to 15 degrees on the right is thus represented on both sides of the brain and is mapped on each side by a disproportionately large number of neurons; this is the area of the barn owl's greatest acuity in locating a sound source.

evokes the maximum response. Changing the difference away from this value causes the response to decrease and finally to cease. The pattern is not affected by the changes in the average sound level; it depends only on the difference in intensity between the ears.

The evidence derived from probing the brain of the barn owl shows that the map of space in the MLD is created by the same cues the owl was known to rely on in finding the azimuth and elevation of a sound: ongoing time disparities and the interaural spectrum. By means of an exquisitely precise transformation the auditory system converts these cues into spatial information. The arrangement of the cells of the map implies that neighboring neurons respond to cues that are only very slightly different. Moreover, the order of the excitatory fields in the map follows the continuity of space. How the brain of the owl is able to achieve such precise connections presents an intriguing problem for further investigation.

SECTION

II

BEHAVIORAL FUNCTION

. . .

Introduction

The past two decades have seen a renaissance of interest in evolutionary theory and its application to natural systems. In part, this was due to the realization that new quantitative tools (simple mathematical models and computer simulations), many of them stolen from economics, can be used to explore arguments efficiently. Verbally expressed ideas that tended to become increasingly murky (or bog down completely) when multiple layers of complexity were added to them can now be pared down to their bare essentials and built up rigorously, step by step.

Although none of the chapters in this section involves the construction or explicit use of such mathematics, the two central theoretical topics under investigation—sexual selection and evolution of altruism—owe much of their current favor to such models. One obvious advantage of mathematical modeling is that many ideas and variations of ideas can be examined for their inherent plausibility before undertaking the labor-intensive and expensive step of testing in the field. In this section we see the fruits of theory-empiricism interactions in the form of four field studies focusing on key theoretical predictions about avian social behavior.

Sexual selection was one of Darwin's many conceptual breakthroughs, wherein he realized that traits may rise and fall solely as a result of their influence on an individual's ability to acquire mating partners. This insight helped him explain a number of awkward features, like the ludicrous tails of male peafowl, that looked like come-and-get-it signals to predators. He reasoned that great success by brightly colored males in mating could easily compensate for their reduced longevity. Darwin saw that this accounted for the counter-intuitive fact that male birds tend to be the more brightly colored sex and that flamboyance was most extreme in animal groups whose males provided little or no parental care. Those males would be under selection pressures to be increasingly sexy, while their conspecific females, with all incubation and other parental duties, would be better off with drab, camouflaged plumages.

A fraction of a population's males may obtain most of the matings via either or both of two processes. The "haves" may be successful because of their ability to outcompete the "have-nots" for mates (and Darwin was particularly impressed with male fighting and intimidation as central to this rivalry). Alternatively, females may simply prefer certain males over others, with an "aesthetic sense" that consequently requires some explanation. Furthermore, these two processes might interact if females benefit by choosing dominant individuals as mates, which have proven themselves to be winners at male-male competitions.

Like peacocks, two of the most extraordinary bird groups in terms of sexual differences are the birds-of-paradise (with exotic male feathering that almost invites disbelief) and the bowerbirds (with fantastic male-constructed courtship structures). These two groups have impressed collectors and field naturalists for centuries, but have received detailed quantitative study only very recently.

The first of the two chapters on sexual selection is Chapter 6, "Sexual Selection in Bowerbirds," by Gerald Borgia, which summarizes the research on satin bowerbirds and addresses the question of why some males do so much better than others in attracting females. Several competing hypotheses are laid out and evaluated as to determine how each fares in explaining the observable facts and experimental results. Chapter 7, "The Birds of Paradise," by Bruce M. Beehler deals mainly with the problem of why polygyny evolves in some species but not in others. The explanation that emerges shows the growing bridge between behavior and ecology, in this case how simple-looking dietary shifts can have far-reaching repercussions throughout all aspects of species' behavior and morphology.

Studies of cooperative breeding address another very basic evolutionary question that lies at the root of sociality. As Richard Dawkins put it, if natural selection can be characterized as "survival of the fittest," then we need to know the fittest *what*? The title of his book *The Selfish Gene* makes clear that Dawkins champions genes as the fundamental unit of selection, although he readily concedes that whole bodies (individuals) are fundamental enough when behavior is the phenomenon of interest. Individuals that act in a way to promote their own lifetime reproductive success tend to leave more offspring (more copies of their genes) and thus selection is expected to favor self-promoting kinds of behavior. It follows that selection should relentlessly penalize self-sacrificing, or altruistic, behavior. From this type of argument, interest naturally fixes on any apparent exceptions to the rule, namely altruistic-looking behaviors such as alarm-calling (where the warner presumably renders itself more conspicuous and more vulnerable on behalf of the warnees) and cooperative breeding (where the "helper" apparently forgoes breeding and contributes substantially to the reproductive success of other individuals).

A key concept underlying the cooperative breeding puzzle is W. D. Hamilton's idea of inclusive fitness, which recognizes the fact that close kin carry identical copies of many genes because of their common ancestry. Your full sibling, for example, should have on average half of the same genes that you do; your cousin, an eighth, and so on. From this it follows that assisting close relatives reproduce successfully is an alternative way of contributing to the gene pool (that is, genes underlying any predilection to perform such services can spread).

This concept would be simple enough if that were all there was to it, but in fact we see a complex dynamic between these alternative routes to Darwinian fitness. While altruistic-looking behavior can spread if directed selectively toward close kin, it spreads even more efficiently if it is actually selfishness in disguise — after all, your genetic commonality with yourself is 100 percent, not 50 percent. Consequently, studies of cooperative nesting in birds must attend closely to the key issue of which players actually reap the benefits of helping.

This dynamic is shown beautifully in Chapter 8, "Cooperative Breeding in the Acorn Woodpecker," by Peter D. Stacey and Walter D. Koenig. When ecological constraints do not permit individuals the luxury of breeding, coalitions form readily between same-sex kin and gains in fitness are achieved via the indirect route of helping and sharing. This behavior can be very complicated because more than a single pair of breeders commonly share the group's only nest. When an opportunity arises for an individual to practice direct self-promotion (for example, to vie for a chance to breed), cooperation quickly goes out the window and even siblings may become serious rivals.

Chapter 9, "The Cooperative Breeding Behavior of the Green Woodhoopoes," by J. David Ligon and Sandra H. Ligon, describes how breeding of woodhoopoes differs in several respects from that of acorn woodpeckers. Most strikingly, woodhoopoes have only one breeding pair per territorial group, so the options of helpers seldom include personal reproduction while living at home. On the other hand, predation is both intense and sex-biased, so reproductive opportunities occur with some regularity: the helper's chance of acquiring coveted breeding status is not trivial if it can but stay alive. Beyond mere patience, however, woodhoopoe juveniles actively participate in raising younger siblings and seem to recruit same-sex siblings into invasion teams that can assist the dominant elder ex-helper in its bid to take over the rare territorial vacancy. All this adds up to a pattern of reciprocal altruism, exchanges of largesse, that can yield net benefits to the individuals because of rapidly shifting needs and opportunities. Because the initial donor eventually receives a full payback with interest, its self-sacrifice is only temporary and the behavior is actually self-promoting in the long run.

In sum, Chapters 8 and 9 reveal some of the circuitous ways in which selfishness, promoted inexorably by natural selection, may account for much of the most generous-looking social behavior known. This is not to say that the indirect-fitness contributions, that is, fitness gains made via the success of nondescendant kin per se, are trivial in these particular cases, much less in other cooperatively breeding species, but it is clear that multiple evolutionary pathways are likely to be involved.

Sexual Selection in Bowerbirds

*The bower, or mating site, of these extraordinary birds of Australia
and New Guinea is the center of intense competition among males.
The female's mating choice is based on its architectural adornment.*

· · ·

Gerald Borgia
June, 1986

Naturalists have long been captivated by the complex and highly elaborate bowers of bowerbirds. Charles Darwin approvingly quoted the contemporaneous assessment of the ornithologist John Gould: "These highly decorated halls of [bowerbird] assembly must be regarded as the most wonderful instances of bird-architecture yet discovered." The interest of the two men is not hard to understand. The bowers of bowerbirds are found from dense forests to open grasslands in New Guinea and Australia, and they have no parallels in the animal world.

There are 18 bowerbird species, and the males of 14 of them decorate clearings or build bowers. There are several kinds of bower (see Figure 6.1). The avenue bowers are formed of two vertical walls of sticks built on a broad platform base. In most species the end of the avenue opens onto a display area where decorations are exhibited. The maypole bowers are made of sticks woven around a sapling or a fern and surrounded by a circular raised court. Two species build a massive hutlike structure about 1.5 meters high around the maypole. The structure encloses a domed runway that opens onto a cleared exhibition area. The golden bowerbird places sticks on adjacent saplings joined by a cross branch, which is used as a display perch. Two other species clear and decorate display courts, but they do not build bowers. The males of one such species build a mat of ferns decorated with shells, and they drape lichen over nearby trees. The males of the other species clear a court on the forest floor decorated only with large leaves.

The decorations associated with the bowers vary greatly among species and include naturally found objects: snail shells, pebbles, feathers, insect parts and bits of bone. Near human settlements the decorations include man-made objects that match the color of natural objects: coins, clothespins, plastic bottle tops, pieces of glass, jewelry, paper, teaspoons, nails, screws, thimbles and the like. The stealing habits of the great gray bowerbird are well known to the aboriginals of northern Australia. Fathers-in-law in that society are notorious for taking whatever they want, and the aboriginal word for father-in-law, *juwara*, is also applied to the bowerbird.

Darwin's fascination with bowerbirds went far deeper than mere appreciation of their coloring, playfulness and architectural skill. When bowers were first noticed by Western observers, they were thought to be nests. As early as 1865, however,

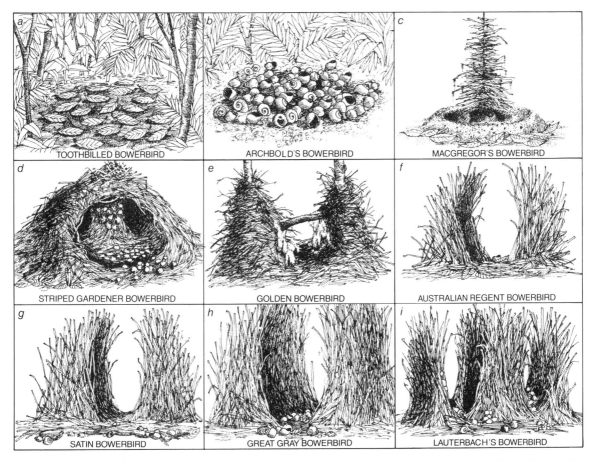

Figure 6.1 MALE BOWERBIRD CONSTRUCTIONS vary from simply ornamented clearings (*a*) to elaborately decorated bowers. The mat bower (*b*) is a mat of lichens decorated with piles of snail shells. The maypole bower (*c–e*) is built of sticks woven around a central pole and is surrounded by a circular raised court. Two species build a large hut over the maypole and pile decorations on a court near the entrance (*d*). One species (*e*) piles sticks on adjacent saplings joined with a cross branch that functions as a display perch. The avenue bower (*f–i*) is built of walls of sticks that enclose the avenue, which opens onto a platform. Lauterbach's bowerbird (*i*) builds an avenue bower with a second set of walls.

Gould carefully observed that bowers were sites for sexual display and mating. Darwin, who was familiar with Gould's work, discussed bower building in his book *The Descent of Man and Selection in Relation to Sex*. Apart from its use as a stage, however, the function of the bower had remained obscure until the recent resurgence of interest in sexual selection. Bowerbirds and their bowers afford a unique opportunity to evaluate competing theories of sexual selection. With the new theoretical tools biologists are now beginning to understand the evolution of one of the most extraordinary behaviors in animals, and

it is possible to give a much fuller answer to the question: Why do bowerbirds build bowers?

Darwin viewed sexual selection as a process separate from natural selection because the main selective forces in sexual selection are social rather than environmental. Indeed, exaggerated male display characteristics, such as the ornate plumes, calls and dances seen in a variety of avian groups, cannot be reasonably explained without considering the effects of social interactions. Darwin described two elements in sexual selection: male competition for females and female choice of males. Male competition is evident in many species and its importance is generally accepted. The role of female choice, however, remains a hotly debated issue.

The codiscoverer of the process of natural selection, Alfred Russel Wallace, disputed the importance of female choice in animal courtship. He doubted that females of nonhuman species had the mental capacity to differentiate among males. The capacity of females to choose is no longer at issue, but the criteria on which the choice might be based and its relation to male competition must still be resolved. In a wide variety of animals as different as scorpionflies and mockingbirds male contributions of food or other materials appear to be the basis of female choice. The males of most bird species take an active role in parental care, and the likelihood of male assistance as perceived by the female seems to be critical in determining the pairing patterns. Male contributions are generally labor-intensive, and in such species the females benefit by choosing unpaired males as mates. The result is that among birds most pairings are monogamous.

For a few avian species, however, such as the prairie chicken, the cock-of-the-rock, the peafowl and bowerbirds, the males offer no direct assistance to the females and provide them only with sperm. In such species the males have evolved extreme characteristics of sexual display, and their effectiveness in attracting mates varies widely among individuals. Yet in spite of the lack of tangible inducement, the females of such species show a strong preference for particular males. There is much discussion about why such preferences exist and how they might be related to the evolution of elaborate displays in males.

There are several divergent views about how sexual selection functions when males contribute only sperm to females. The two most widely discussed general hypotheses are known as the good-genes models and the runaway models. In both kinds of models the female actively chooses her mate, and both models give a plausible explanation for the evolution of bowers. Each of them, however, depends on unproved assumptions, and neither has been shown to have operated in a natural population. Two other models, the proximate-benefit model and the passive-choice model, have been much less discussed, but they may turn out to be important in understanding the evolution of male display.

The hypotheses collectively known as good-genes

models stress that a female that discriminates among her potential mates can enhance the overall health and vigor of her offspring. Thus according to the model, male displays evolve because they provide females with information about the relative quality of a male as a sire. But to what aspects of the male should a discriminating female attend?

Richard D. Howard of Purdue University, Tim Halliday of the Open University and several other biologists have suggested females favor older males over younger ones. The older males have presumably demonstrated their hardiness simply by having lived to an advanced age. For example, suppose older males tend to carry heritable traits promoting survival, such as disease resistance or predator avoidance. Then all else being equal, if the good-genes hypotheses are correct, the female should prefer the older males as mates. How is she to make that choice? One plausible strategy might be to attend to the more elaborate display characteristics among her suitors on the assumption that the more practiced and elaborate the male's display, the older the male. Thus the female's search for an older mate might explain the elaboration of male display characteristics.

Females might also seek the genetic enhancement of their offspring by favoring active males able to court vigorously or by favoring males with bright plumage, which could indicate male health and disease resistance. Hence the sexual-display patterns of the males, including their exaggerated plumage and decorated bowers, may have evolved to provide information to females about heritable, fitness-enhancing traits. Complex traits, such as the male's overall vigor, may summarize the effects of genes throughout the genome. Such traits could be quite useful to the female in choosing a genetically superior partner.

In many animal species males compete with one another for access to females. Displays that show the dominance of a male in aggressive encounters may reliably indicate his superior fitness relative to other nearby males. I have therefore suggested females may prefer to choose among males that give ostentatious displays because the females are seeking dominant males. Males giving such displays without harassment must be dominant because subordinate males attempting to give the same display would be challenged by higher-ranking males.

If mates are selected according to display patterns alone, one would expect that all mature males would have the plumage needed for a display but that only some of them would earn the opportunity to show it off. There is an analogue to this prediction for the decoration of bowers. Male bowerbirds continually attempt to destroy other bowers and steal their decorations, and that behavior is analogous to the competition for the opportunity to display plumage. Together with Stephen G. and Melinda A. Pruett-Jones of the University of California at San Diego, I have proposed that a male's ability to maintain a decorated bower of high quality may serve to indicate to the female his relative quality as a sire.

Amos Zahavi of the University of Tel Aviv has suggested the female might gain genetic benefits for her offspring if she favors a so-called handicapped male: a male displaying debilitating physical characteristics or behaviors. According to Zahavi, the elaborate sexual displays are genetically inherited handicaps. The female allegedly gains genetic benefits by choosing a male with a handicap because such a male has survived in spite of a highly disadvantageous trait.

The model has met with much criticism. The most significant flaw is its reliance on the environment to select only the fittest of the handicapped males. If the females are to find the best sires in the population, there must be a very high rate of male mortality attributable to the handicap. Otherwise the overall genetic superiority of the handicapped survivors in the population could not be guaranteed. The model's requirement of high mortality among the sons of handicapped males makes it unlikely that females preference for superior males could evolve according to such a scheme.

The runaway model for sexual selection was proposed by Ronald Fisher in 1930 to explain the evolution of exaggerated characteristics found only in males. He was the first to note that the pattern of female choice could be self-reinforcing: it could cause its own spread among females and the spread of the male-display characteristics on which it is based.

To illustrate the runaway process suppose there is a population in which there are two kinds of male and two kinds of female. The males differ in the presence or absence of a display characteristic, such as a red tail feather. The females differ in that some of them (the "choosers") mate only with males having a red tail feather, whereas others (the "nonchoosers") do not distinguish among males on the basis of tail-feather color.

In such a population the males having a red tail feather can mate with both chooser and nonchooser females. In contrast, males without a red tail feather can mate only with nonchooser females. Thus redtailed males have more opportunities to mate and produce a greater proportion of progeny than males without a red tail feather; the proportion of redtailed males in the population thereby increases. The sons of red-tailed males and chooser females carry a greater-than-random proportion of the genes that lead to choosing behavior in females. As red-tailed males mate more often, the proportion of the traits specifying a female preference for redtailed males also increases.

Once a pattern of female choice is established, there can be continued selection for more exaggerated male traits. The outcome of the runaway process depends strongly on the choice pattern. If females consistently prefer males having extreme characteristics, such male characteristics are expected to evolve. If, however, females prefer males with a less extreme trait, such as a single red tail feather, males in the population with one red tail feather will tend to predominate over males with an all-red tail. The survival costs of sexual display can affect the process. For example, as extreme characteristics develop in males, new female-choice patterns may arise that cause a reverse runaway toward less extreme male display characteristics.

Very little is known about how new femalechoice patterns arise. Moreover, when they do arise, it is not clear whether they favor only a slight enhancement of male characteristics or an extreme development. Several recent models of runaway selection conclude that female choice can be completely arbitrary and so can give rise to the evolution of arbitrary male characteristics.

The good-genes models and the runaway models thus make differing predictions about how extreme display traits are established. According to the good-genes model, natural selection should favor female preferences for traits that indicate differences in male qualities as sires. In the runaway models such an outcome is not necessarily expected. A runaway selection may not enhance the vigor of the offspring, and it might even promote traits that reduce their fitness.

The extent to which traits that reduce fitness might evolve is still a matter of debate. It would seem more likely that if different kinds of female preference were expressed in a population, the kinds tending to enhance fitness would have an evolutionary advantage over the ones that did not. The extent to which traits enhancing fitness win out over less advantageous ones must depend on at least two factors: the ease with which established traits can be replaced and the frequency with which competing patterns of choice are found in a population.

Comparing the good-genes models with the runaway models is further complicated because the models need not be mutually exclusive. For example, if the initial female choice depended on the effects of good genes, it could lead to the selection of female-preference traits that are self-reinforcing. It is therefore unrealistic to expect the predictions of these models to be easily distinguishable in natural populations. Nevertheless, if the controversy is to be resolved, it will only be through studies of natural populations. One approach is to determine whether or not the patterns of male display are really arbitrary. Alternatively, one could show that they do indeed indicate the relative quality of the males as sires.

I noted above that there are at least two more possible explanations for the evolution of bower-building behavior. According to the proximate-benefit model, there may be immediate benefits for the female that chooses a male giving an extravagant display. For example, bright male plumage may better enable females to detect parasites and so avoid contact with males likely to transmit the parasites. Females that discriminate against infested males gain an obvious, immediate benefit over less choosy females. Other kinds of immediate gain might also arise for the discriminating female. A male giving a display without interference from other males might be able to offer superior protection from predators, and males with elaborate plumage are more likely to be mature and so carry viable sperm.

Finally, some biologists have suggested female choice has had little to do with the evolution of exaggerated male display. According to the passive-display model, put forward by Malte Andersson of the University of Göteborg and Geoffrey A. Parker of the University of Liverpool, elaborate male displays may have evolved as advertising devices. Males with the most ostentatious displays are more readily visible to females and so such males have more chances to mate. Gains from the additional matings repay the males for the extra cost of surviving while giving an extravagant display. The model does not require that females exhibit any active preference for males with larger displays. If a female simultaneously encounters two males that differ in the extent to which their displays are elaborated, the model predicts the size of the displays should not influence her mating decision. For bowerbirds this model implies that the bower is a device for advertising to females the presence of a courting male and that males with larger, better decorated bowers are sexually successful because they are more often found by females.

Although substantial efforts in constructing theoretical models have sharpened the questions one would like to answer about how elaborate displays have evolved, there have been few attempts to test the models in natural populations. In part the lack of testing is a result of the difficulty of finding observable subjects with appropriate characteristics. The special characteristics of the bowerbird's display make it possible to study female preferences in detail. I have already mentioned that male bowerbirds give no material assistance to females or to their young, and the females are free to choose among males from widely separated display sites. One can therefore assume that biases observed in female preferences for males are related to characteristics of the male's display, including his bower, plumage and behavior in the presence of the female.

Furthermore, there may be a functional equivalence between the decorations on the bower and brightly colored plumage that may make it possible to manipulate the general display patterns experimentally. More than two decades ago E. Thomas Gilliard [see "The Evolution of Bowerbirds," by E. Thomas Gilliard; SCIENTIFIC AMERICAN, August, 1963] suggested there is an inverse correlation between the degree of plumage elaboration in males and the size and degree of the decoration of the bowers. Thus, Gilliard noted, the decorated bowers may play the same role in courtship as showy displays of plumage; in fact, he suggested, bowers are a kind of displaced plumage that allow the animals building them to dispense with bright coloration. The development is called the transfer effect (see Figure 6.2). If it is real, it suggests the same forces shaped both the evolution of bower building and decorating behavior and the displays of showy plumage.

Unlike plumage, the bower and its decorations can be easily manipulated and quantified with no direct effect on the bird. Matings take place at the bower, and so cameras monitoring the bowers can

Figure 6.2 TRANSFER EFFECT is illustrated by the bowerbirds and their bowers. The male Australian regent bowerbird (*Sericulus chrysocephalus*), which displays bright plumage, builds a bower of indifferent structural quality and makes little attempt to decorate it (*left*). In contrast, the male great gray bowerbird (*Chlamydera nu-*

chalis), a bird of dull color, builds an elaborate bower that is richly decorated with shells, flower petals and a variety of manmade objects (*right*). Decorated bowers may have the same function for the male as colorful plumage displays, and any hypotheses that explain the evolution of the bower may also explain the evolution of colorful plumage.

record the choice of a mate by the females and the mating success of the males. The observations can be compared with the quality of the bower and the elaborateness of its decoration. Finally, male behavior that can be observed near the bowers is an important indicator of how competition can distinguish among potential suitors. Males often steal bower decorations and destroy the bowers of other males, and the patterns of such aggressive behavior can be compared with the quality of a male's display and his success in finding a mate.

In 1980 I began an intensive study of the satin bowerbird (*Ptilonorhynchus violaceus*) in eastern Australia. J. M. Marshall, Reta Vellenga and Richard Donaghey, who was then at Monash University, had done important early work, and their efforts made the life history of satins the best documented of any bowerbird species. Donaghey allowed me to take over a study population with which he had worked, so that the histories and identities of some males had already been known for four years. At that time the details of female choice were not known for any bowerbird species.

The satin bowerbird ranges along the southern and central east coast of Australia. My research site is in a valley formed by Wallaby Creek in Beaury State Forest of New South Wales (see Figure 6.3). Vellenga had shown that the male is slightly larger than the female and gets his satiny blue plumage in his sixth year. The coloration of adult males differs markedly from the dark green back and spotted yellow-white underside of females and juveniles. The male builds bowers made of sticks on courts cleared on the ground (see Figure 6.4).

The north end of the bower faces the sun at midday and opens onto a display platform, which the mature male covers with bright yellow straw and yellow leaves. A variety of decorations are laid on the platform, including blue parrot feathers, blue and yellow blossoms, insect parts, in particular the outer coverings of cicadas, and other natural objects. Large objects including the shells of land snails are arranged at the outer edge of the platform, and feathers are distributed evenly over the yellow platform between its outer edge and the avenue of the bower. Small objects held by the male in

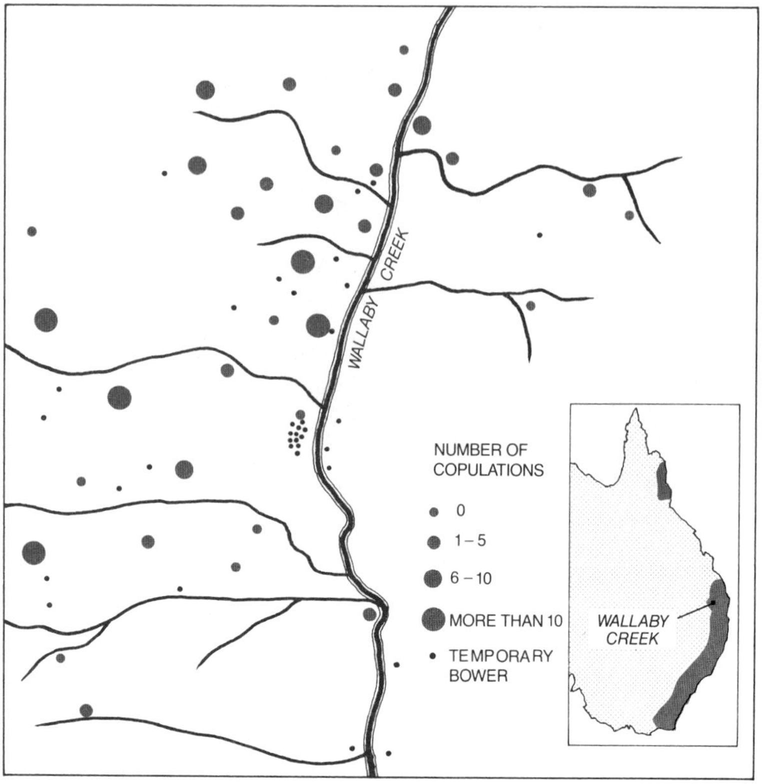

Figure 6.3 AUTHOR'S STUDY AREA is in a valley formed by Wallaby Creek in the Beaury State Forest of New South Wales. The inset map shows the location of the site as well as the range of two subspecies of satin bowerbird: *P. violaceus violaceus (color)* and *P. violaceus minor (gray)*. Rain forest predominates in the low areas, along creeks and on the eastern side of the ridges west of Wallaby Creek. Sites of established male bowers of the study area are shown as colored circles; the diameter of the circles indicates the number of matings per season for the bower's owner. Sites of temporary bowers, built by younger males and destroyed soon after, are shown as black circles.

his mouth during courtship are found in a small pile near the bower.

The carpet of yellow straw and leaves on the bower platform creates a bright glow that is particularly noticeable at bowers in forests. The male prunes the leaves above the platform, apparently to allow sunlight to illuminate the platform. The display of shiny blue objects, which are relatively uncommon, and their placement on a yellow background suggests an attempt to give an unambiguous and highly visible signal.

In order to study the behavior of individuals we capture bowerbirds flocking in open pastures. Each bird is fitted with a unique color-band combination, measured and assessed for plumage, the color of its beak and legs, scratches or other evidence of fighting and external parasites. We assign the bird to an age category according to the colors of its plumage and its beak.

My volunteer assistants, graduate students and I record the behavior of males at feeding sites, noting

the number of times a male is attacked and how often he attacks others. At the bowers we observe the birds from hides, and we also continuously monitor activity at bowers with remote-control super-8 motion-picture cameras throughout the mating season. The cameras expose a frame every two seconds as long as there is a break in an invisible beam of infrared light passing through the bower. This record enabled us to identify individual visitors to the bowers and to note their activities,

Figure 6.4 SATIN BOWERBIRD is shown standing on his bower platform. The deep, iridescent satiny blue coloration identifies the bird as a mature adult male. The bower is built by the male on a court cleared on the ground. It is made of sticks woven into two vertical walls that enclose an avenue on the surface of the broad platform base. The decorated platform serves as a stage for the display of the male and as a mating site. The structural quality of the bower and its decorative embellishments are major factors in the female's choice of a mate, and so the bower is the focus of the competition among male bowerbirds for mates.

including the destruction of the bower, the stealing of its decorations, courtship and copulation. We also made daily records of the bower quality and the movement of marked decorations at more than 33 bowers in the course of the study (see Figure 6.5).

At Wallaby Creek mating begins in early November and continues until late December. In mid-October males become active around the bowers: rebuilding of the bowers at permanent sites is completed, and intensive decorating is begun (see Figure 6.6). Young males visit the bowers and the bower owners often display to them. By early November, however, the bower owners become less tolerant of male visitors; they spend more time near the bower and engage more actively in destroying nearby bowers and in stealing their decorations. Females overwintering at the southern end of the valley start moving north toward the highest concentration of male bowers. In mid-November the matings begin. The males then spend most of their time perched in trees near the bower, calling frequently and moving down to the bower to display to females, to protect it from marauders, to build it or to "paint" it with the saliva generated in chewing bits of vegetation. Matings peak at the end of November and are mostly finished by mid-December.

When a female visits the bower, the male begins his display, often holding a decoration in his beak. He faces the female while he stands on the platform. He gives a whirring call while prancing, fluffing up his feathers and flapping his wings to the beat of the call. Calls are punctuated with periods of silence, quiet chortling, buzzing or mimicry of other birds. The female's initial response is to enter the bower and "taste," or nip, at a few sticks. Then she intently watches the courtship. If she is ready to copulate, she crouches and tilts forward. The male immediately mounts her. At any stage she may leave, thereby ending the courtship. Typically a female mates only once. She later lays two eggs in a nest that is usually outside the area defended by her mate. The hatching of the eggs coincides with the emergence of large numbers of cicadas.

Established males destroy bowers other males may try to build nearby, but during the mating season young males can establish temporary bowers at sites removed from the permanent ones. There is intense courtship at the temporary sites among males that appear to be practicing and learning the display. Such males also visit permanent bowers. If the owner is not present, the visitor may paint the bower, attempt a display or court a visiting female. During courtship by a bower owner other males may hide in the surrounding vegetation and then try to interrupt or displace a copulating male.

Most hypotheses about the evolution of elaborate displays assume the quality of the display affects the female's willingness to mate. For satin bowerbirds mating success varies widely among males: we observed one male that mated with 33 females during the season, whereas many other males did not mate at all. When we ranked bowers for quality, we found a strong positive correlation with male mating success. Neat and well-built bowers with symmetrical walls, fine, densely packed sticks and a highly sculptured appearance were owned by particularly successful males.

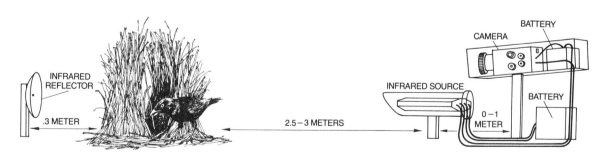

INFRARED REFLECTOR

.3 METER

2.5 – 3 METERS

BATTERY

CAMERA

INFRARED SOURCE

BATTERY

0 – 1 METER

Figure 6.5 EXPERIMENTAL MONITORING SYSTEM employed by the author in his study of the satin bowerbird is diagrammed schematically. An infrared beam, invisible to the bowerbird, is projected through the avenue of the bower to a reflector. When the beam is interrupted, a super-8 motion-picture camera exposes one frame every two seconds. Birds were also observed from blinds. The system enabled the author to monitor the behavior and identity of bower owners and visitors at 33 bowers for the 50-day mating season.

Figure 6.6 BREEDING BEHAVIOR of the satin bowerbird begins with the male building his bower (*top left*). When a female visits the bower, the male moves down from a perch in a nearby tree and begins courtship (*top right*). After watching the display the female may crouch and tilt forward, and the male immediately mounts her (*middle left*). Other males may compete by destroying the bower in the absence of the owner (*middle right*) or by stealing its decorations. A competing male may even try to interrupt or displace a male during copulation (*bottom left*). After successful copulation the female rears her young in a nest that is separate from the bower (*bottom right*).

To test the importance of bower decorations in the female's choice of a mate we removed the decorations from the bowers of a randomly selected group of males. We then compared their success in mating with the success of a control group we did not disturb. As we had predicted, the decorations do indeed influence mating: we found a significantly higher rate of mating in the control group than we did in the experimental group. Moreover, the number of decorations on the platforms of undisturbed bowers—particularly the number of feathers—was correlated with male mating success in each year of the study.

The discovery that decorations are important in male display led me to a study of decoration stealing. We found that blue feathers are stolen at a much higher rate than other decorations in proportion to their frequency on the bower platforms, and that (because they are rare in the habitat) stealing is the principal means of obtaining them. By monitoring the number of feathers on bower platforms throughout the mating season we discovered that the number on the platforms of successful males peaks at the height of the mating season. At the same time the number of feathers on the platforms of unsuccessful males is reduced.

In another experiment we introduced individually marked blue feathers to the bowers of a group of males, and we added no feathers to the bowers of a control group. We then reversed the treatments of the experimental and the control groups, and we recorded the movements of the feathers in each instance. Yet whatever the initial placement of the feathers, the same males, namely the most active thieves, tended to accumulate the feathers on their bower platforms. The result is strong evidence that stealing is the most important factor in the final number of feathers displayed on the platforms. Because the quality of the bower decoration affects the success of the male in mating, feather stealing appears to affect mating success. More dominant males tend to be more successful at feather stealing.

We also found that the average quality of a male's display depends on the frequency with which it is destroyed by marauding visitors. The more frequent the destruction, the lower the overall quality of the bower. The pattern of destruction can indicate male dominance to the female because females generally limit their bower visits to a small area. If a bower is maintained in relatively good condition, it can serve as a signal to the female that the owner can defend it from attack and destroy the bowers of his neighbors as well. The behavior patterns we found for bower decorating are consistent with the patterns of bower destruction: the bowers of older, dominant males are destroyed less often than those of younger, subordinate males.

How do such observations conform with the models for the evolution of selection patterns I have described above? There are several lines of evidence to suggest females favor dominant males. They choose mates that are able to keep bowers in good repair and well decorated. Such males tend to be the ones that are dominant at feeding sites; their decorative preferences are for objects scarce in the habitat and prized by other males. The age of the male also seems important. Older males maintain bowers of better quality and decorate them more elaborately, and they are more successful in protecting their bowers from destruction. Moreover, older males give more refined courtship calls.

It may be that both male dominance and age are important to the female's choice. Bower building appears to take some practice, but only the most aggressive young males are able to practice building and decorating bowers in the face of repeated destruction by other males. A female that chooses an older, established male with a well-built, well-decorated bower and a refined courtship call has evidence that her prospective mate not only has been able to survive to a relatively old age but also has been able to do it while learning to build and maintain a high-quality bower under the rigors of male competition.

The remaining good-genes model, the handicap, is not supported by the behavior of bowerbirds. The model predicts that large differences in male mortality are associated with differences in the quality of male displays. That is not what one finds. The mortality of displaying males is low, and it appears to be independent of the quality of male displays.

The runaway model can lead to a large number of possible outcomes. Versions of the model that suggest arbitrary outcomes yield no prediction about the kinds of traits that should evolve; hence they cannot easily be falsified. In some cases an arbitrary choice resulting from runaway selection might give rise to the same behavior as a choice made on the basis of male dominance. For example, if females had evolved a tendency to favor males with well-built, well-decorated bowers, males able to steal decorations and destroy bowers, or in other words the dominant males, would thereby be selected.

The existence of patterns consistent with other models weakens the case for the runaway models, particularly if the patterns can be shown to recur in various species. In satin bowerbirds we found that males favored scarce decorations. If a similar pattern in other species were found, it would support the suggestion that the female assessment of male dominance is important in the choice of a mate.

Can passive choice explain the patterns of mate selection we observed? Probably not. We found that sexually successful males tend to receive more female visitors to their bowers than less successful males, and they mate with a greater proportion of the females. Generally females visit several bowers before mating, and their choice is correlated with the overall mating success of the males. In contrast, passive choice seems unlikely. Bowers are on the ground and often under cover; the call of the bower owner from his nearby perch is much more noticeable to a naive female than the bower itself. Furthermore, females are long-lived and therefore probably familiar with all the bower sites in the area they search; it seems unlikely that mere prominence would affect the choice of such females.

There may be some proximate benefit from the bower for the female as a protection from intruding males. We often saw males trying to interrupt matings by bower owners. Usually the owner chased the intruder away and the female remained in his bower, although occasionally copulation was interrupted. When females were on the ground outside a bower, however, they seemed to be uneasy and commonly flew away if intruders came near.

To summarize, the evidence available suggests female satin bowerbirds actively differentiate among males according to the quality of their displays. The females may also choose sires according to the decorations associated with the bowers. The protection hypothesis offers an alternative explanation for the evolution of bower building, but it does not explain why the bowers are decorated. The runaway model cannot be excluded as a possible alternative explanation.

The work on satin bowerbirds is a first step in understanding the evolution of exaggerated characteristics, and we have established the plausibility of several models of that evolution. Nevertheless, studies of mating choice in other bowerbirds will be necessary if one is to explain why bowerbirds build bowers. To what extent do other species show a preference for rare decorations? Do they steal the decorations of other bowerbirds and destroy their bowers? How do males learn to give their displays? Will the findings suggest the same causal relations between male behavior and female choice? Why are bowers built by some species but not by others? Such questions will surely be resolved by further patient observation.

The Birds of Paradise

Diet has a major influence on the social and sexual behavior of these tropical birds. It explains why some species are monogamous and why others are highly promiscuous.

. . .

Bruce M. Beehler
December, 1989

The birds of paradise, denizens of a tropical realm far removed from the museums and laboratories of the Western world, have been the focus of scientific curiosity for many years. In 1871 Charles Darwin referred to them in his book *The Descent of Man and Selection in Relation to Sex*, noting that "when we behold a male bird elaborately displaying his graceful plumes or splendid colors before the female . . . it is impossible to doubt that she admires the beauty of her male partner."

Although nearly 120 years have passed since Darwin first remarked on the plumage of the male birds of paradise, the evolution of specialized plumes and other aspects of the birds' reproductive behavior continues to interest evolutionary biologists. Within the past 10 years a new generation of investigators—armed with fresh insights from sociobiology and behavioral ecology—have learned much about social organization in the birds of paradise.

Thane K. Pratt of the U.S. Fish and Wildlife Service in Hawaii, Clifford B. and Dawn W. Frith of Wildlife Conservation International, Stephen G. and Melinda Pruett-Jones of the University of Chicago and I have collectively invested more than a decade in the field studying the lives of these birds. Thousands of hours of field observation have

shown that the diversity of social and sexual behaviors in this remarkable avian family has an ecological basis. In addition, it is now known that the birds, which feed mostly on fruit, play a key role in maintaining and regenerating the Papuan rain forest.

My own work on the birds of paradise began in 1975 with a 15-month sojourn in Papua New Guinea. Although 14 years have passed since I first set foot on the island, an early foray into the forest remains sharply fixed in my mind. Pushing my way through a lush tangle of tropical vegetation in the Upper Watut Valley of eastern New Guinea, I came on a tall tree in which there were eight adult male raggiana birds of paradise—beautiful animals with yellow heads, iridescent green throats and velvety brown breasts. They flashed their long orange plumes as they bowed and displayed to a visiting female, who was perched less than six inches from a male at the center of the group. To me the color, movement and sound of the mating display was—and continues to be—one of nature's most thrilling sights (see Figure 7.1).

Figure 7.1 MALE RAGGIANA bird of paradise displays for a nearby female (not shown) by bending over, head down, and spreading his wings and orange flank plumes. The male raggiana sits on his perch every day for at least six months a year.

There are 42 species in the family Paradisaeidae. Of these, 36 are endemic to New Guinea and its satellite islands, four occupy eastern Australia and two occur as far west as the northern Moluccan Islands of Indonesia (see Figure 7.2). Included in the family are a number of superficially diverse birds: sicklebills, parotias, long-tailed astrapias, heavily plumed paradisaeas and blue-black manucodes. All are robust in build, with powerful legs and feet, yet from species to species there is great variation in color and plumage. In most species the birds are sexually dimorphic, that is, the males and females differ markedly in appearance: the males are often gaudily colored and have long, ornate feathers called nuptial plumes, whereas the females lack the specialized plumes and are always predominantly brown or black. (see Figure 7.3).

The bright colors and courtship displays of the male birds of paradise are testimony to the power of sexual selection, the process by which some individuals win (and others lose) the right to reproduce. Competition among members of one sex (usually male) for access to the other sex (usually female) explains the prominent differences between males and females that are so common in the animal world — for example, the bright-red plumage of the male cardinal versus the dull brown of the female or the enlarged antlers of a buck versus the unadorned head of the doe.

Today much of the debate about sexual selection focuses on the nature of interactions among males as well as those between males and females. Why, for example, is sexual dimorphism so pronounced in some bird-of-paradise species and yet virtually nonexistent in others? Do the remarkable plumes of the males represent badges of dominance within a mating hierarchy, or are they advertisements to a female of a male's physical health and vigor? If the latter is the case, is there evidence that females discriminate between males on the basis of physical characteristics? Although complete answers to these questions are still forthcoming, much has been learned recently about the ecological basis of behavior in the birds of paradise. Before these findings can be presented, however, a review of social organization in the Paradisaeidae is needed.

Not all birds of paradise are sexually dimorphic. Nine of the 42 species in the family are monomorphic: the males and females are virtually identical in appearance. Monomorphic species, such as the trumpet manucode, *Manucodia keraudrenii*, are also monogamous: they form tight pair-bonds and perhaps mate for life. Moreover, both sexes cooperate in raising their young. In contrast, dimorphic species, such as the raggiana bird of paradise, *Paradisaea raggiana*, are polygynous: the males are promiscuous, mating with as many females in one season as they can. After mating, the females receive no additional help from the males and rear their offspring alone.

Why have such different mating strategies evolved within one family? Why, specifically, are some species polygynous and others monogamous? The answer has two parts. The first has to do with the influence of anisogamy, the inequality of male and female sex cells, on polygyny; the second part — as I shall explain later — has to do with the foraging behavior of a species and the nutritional and economic impact of diet on reproductive strategies.

Unlike eggs, sperm are tiny and energetically cheap to produce, and so in many species males can maximize their reproductive output (that is, the number of genes they pass on to the next generation) by mating with as many females as possible. Females, in contrast, produce only a few eggs and so are generally restricted in their reproductive output. The result is a behavioral and morphological dichotomy between the sexes: males tend to be promiscuous and to compete — fiercely at times — for access to females, whereas females tend to be selective, favoring males that possess certain characteristics over others.

Thus, in a monogamous population, where there is little competition because individuals pair-bond, there is little sexual dimorphism. In a polygynous population, where competition is stiff, dimorphism evolves because males with the brightest plumage and most elaborate displays mate with more females and thus pass along more of their genes and traits to the next generation. For a male who is competing with his neighbors, a slight edge — a louder call, more aggressive behavior or more conspicuous plumes — can spell the difference between no matings and multiple matings.

In both monogamous and polygynous populations, females tend to be drably colored. Because they are always in demand by males, they need not look flashy in order to reproduce. In the absence of any reproductive value for bright coloration, natural selection is thought to favor drab females over colorful ones: a cryptic or camouflaged female tending her nest is likely to attract predators searching for eggs or young nestlings.

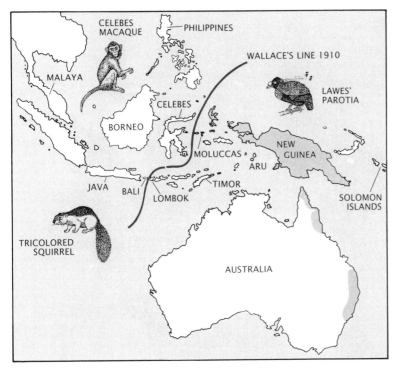

Figure 7.2 BIRDS OF PARADISE are endemic to rain forests in Australia, New Guinea and the Moluccan Islands of Indonesia. In certain respects, the birds occupy the ecological niche held elsewhere by arboreal, fruit-eating mammals. The absence of mammals can be attributed to the presence of a deep-water barrier that separates the islands of Southeast Asia from Australia and New Guinea. The barrier, first recognized by Alfred Russel Wallace in 1863 and redefined by him in 1910, is known as Wallace's Line. To the east of the line are birds of paradise; to the west are placental mammals, including tricolored squirrels and macaques, both of which feed on fruit.

The male raggianas, such as those I saw displaying together in 1975, are typical in many ways of species that are sexually dimorphic and polygynous. Their courtship ritual is elaborate, and because it is performed at the same time by several males clustered together, it is also competitive. A group of courting males, called a lek, assembles at dawn each morning in a display tree. They advertise their presence to females within earshot by calling loudly: *wau, wau, wau, Wau, Wau, Wau, WAU, WAU, WAU, WAU,* with increasing volume and speed. If a female responds by joining them in the display tree, the males will initiate a courtship dance: they raise their orange display plumes, shake their wings and hop from side to side, while continuing to call.

After a brief bout of noisy display behavior, the males become silent and lean upside down from their perches, with their wings thrown forward and the erect orange plumes forming a resplendent fountain of color (see Figure 7.4, *right*). The group holds this pose until the female, who moves silently among them, selects a mate and crouches beside him. The other males watch passively as the chosen male performs a precopulatory dance and then mounts and mates with the female. The female separates from her mate soon thereafter and flies off to her nest, where in a day or two she will lay an egg.

The males of other polygynous species behave similarly, but each species has a unique courtship ritual, which is enhanced by the males' distinctive color and plumage. The male buff-tailed sicklebill, *Epimachus albertisi,* for example, displays by hanging upside down and erecting his plumes to form a circular cape; the magnificent bird of paradise, *Cicinnurus magnificus,* forms a shield of feathers that looks like a high-necked collar and dances like a mechanical toy, tail feathers erect, while clinging to a vertical stem. The Lawes' parotia, *Parotia lawesii,* displays on the forest floor in a most remarkable fashion: while the female observes from above, the male throws his six stiffened head feathers forward like antennae, spreads his plumes into a skirt and hops delicately about (see Figure 7.5).

In some polygynous species, such as the raggiana, the males form mating clusters, or leks, in which they congregate and display for visiting females.

RAGGIANA BIRD OF PARADISE

MAGNIFICENT BIRD OF PARADISE

TRUMPET MANUCODE

BUFF-TAILED SICKLEBILL

Figure 7.3 PHYSICAL DIFFERENTIATION between males and females, or sexual dimorphism, is most pronounced in the raggiana (*upper left*): males are larger than females and festooned with orange flank plumes, which are highly visible when the male displays. Male and female magnificents (*upper right*) are similar in size, but the male is brightly colored and has two unusual feathers, called tail wires, which he flashes when courting. Males and females of the buff-tailed sicklebill (*lower right*) have long, curved bills, with which they extract insects from wood and fruits from inside capsules. Trumpet manucodes (*lower left*) are monomorphic.

The type of lek varies from species to species. Raggianas, for instance, tend to congregate in one tree, whereas the males of Lawes' parotia (see Figure 7.6) form what is known as an exploded lek: instead of clustering tightly together so that each one is only a few feet from his neighbor, they space themselves from a dozen to 100 meters or more apart yet still behave as members of a single mating group. In some species, such as the magnificent, the males are solitary and do not form leks at all.

The presence or absence of leks raises additional questions about the evolution of mating strategies in the Paradisaeidae. In most leks one or two dominant males, usually the ones perched near the middle of the group, mate most often. Only rarely do peripheral males have an opportunity to mate. In a

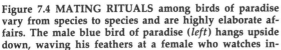

Figure 7.4 MATING RITUALS among birds of paradise vary from species to species and are highly elaborate affairs. The male blue bird of paradise (*left*) hangs upside down, waving his feathers at a female who watches intently from above. The male raggiana (*right*), having completed the upside-down phase of his display, is preparing to mate with the receptive female above him.

lek of the lesser bird of paradise, *Paradisaea minor*, for example, I recorded that a single male carried out 25 of the 26 observed copulations during one month of the mating season. Why, if competition among males is so fierce, do male's cluster together? Might a male's chances of mating not be greater outside a lek than in it? And why do some species form leks, whereas others do not? Answers to all three questions show that lek formation, like polygyny, is related to the foraging ecology of a species.

An initial clue that polygyny might be related to diet came from research conducted in the 1970's by David and Barbara Snow of the British Museum of Natural History. Their studies focused on the relation between mating organization and dietary specialization in two families of birds: the cotingas (Cotingidae) and the manakins (Pipridae), both of which are endemic to the rain forests of Central and South America. Although the cotingas and manakins are only distantly related to the birds

Figure 7.5 FEMALE LAWES' PAROTIA is perched on a branch watching a male display on the ground. The male has fluffed out his flank plumes, forming what looks like a skirt. After dancing he will join the female and attempt to mate with her. Parotia males form exploded leks: they display together but space themselves some distance apart.

of paradise, they exhibit remarkably similar patterns of sexual dimorphism and mating behavior.

The Snows found that most polygynous species are frugivorous (fruit-eating), whereas most monogamous species are predominantly or entirely insectivorous (insect-eating). They also demonstrated that more time and energy is needed to forage for insects than is needed to forage for fruit and reasoned that only by eating fruit would polygynous males have sufficient time in which to pursue their promiscuous mating strategies. Would a similar pattern emerge, I wondered, for the birds of paradise?

As it turns out, the correlation between diet and mating behavior is considerably more complex in the Paradisaeidae than it is in the cotingas and manakins, and if anything, the pattern seen by the Snows is reversed. Research my colleagues and I have done shows that dietary specialization has had a significant influence on the evolution of the Paradisaeidae—promoting species diversity in the family at the same time that it has promoted a shift to polygynous mating behavior (see Figure 7.7).

I have found that the fruits on which the birds of paradise feed fall into two distinct categories: sim-

Figure 7.6 MALE LAWES' PAROTIA looks strikingly different from the female. His vibrant chest colors and the "head wires" (highly modified feathers) at the back of his head are secondary sexual characteristics, which in parotias and other dimorphic species may have evolved in response to competition among males for access to females.

ple, raspberry-like fruits, such as figs, which are plentiful in the forest and provide a ready source of water and carbohydrates, and complex fruits, such as nutmegs, which are large, usually protected by a tough outer capsule and produced in relatively small quantities. Such capsular fruits are often rich in fats or proteins, which makes them valuable commodities for foraging birds.

Pratt and I found that all the birds of paradise and just about every other bird in the rain forest will eat large quantities of figs when they ripen. Because most fig plants produce thousands of small, rapidly ripening fruits in cycles that are both nonannual and asynchronous, figs are available at unpredictable times throughout the year. When a plant does bear fruit, the avian response is spectacular: manucodes, fruit doves, fig parrots, pitohuis, honey eaters and birds of paradise all flock to the plant, gorging themselves on its ripe fruit.

Yet only one bird of paradise, the monogamous trumpet manucode, feeds almost exclusively on figs. Although most other birds of paradise will eat figs opportunistically, polygynous species spend the bulk of their foraging time searching for complex fruits, generally the capsular type. The fruits ripen in small numbers throughout a long fruiting season and are eaten by birds that visit ripening trees on a daily basis. Capsular fruits are more difficult to harvest than figs because they are encased in a tough woody capsule and consist of a single seed (with an edible portion attached), which must be swallowed whole.

It turns out that the ability of certain species to gain access to capsular fruit, which is difficult to reach and difficult to open, is related to their ability to feed on insects. The birds forage on limbs and trunks of trees, hammering and prying away at bark and dead wood to extract bark beetles and other wood borers, much as woodpeckers do. They also feed on insects that hide inside curled leaves by holding the leaves with their feet and pulling the insects out with their bill. The anatomical adaptations, such as a long beak and well-developed claws, that permit such specialized insectivory also make it possible for the birds to feed on capsular fruit.

Fig feeders are rarely seen foraging in the trees that produce capsular fruit, and many of the plants

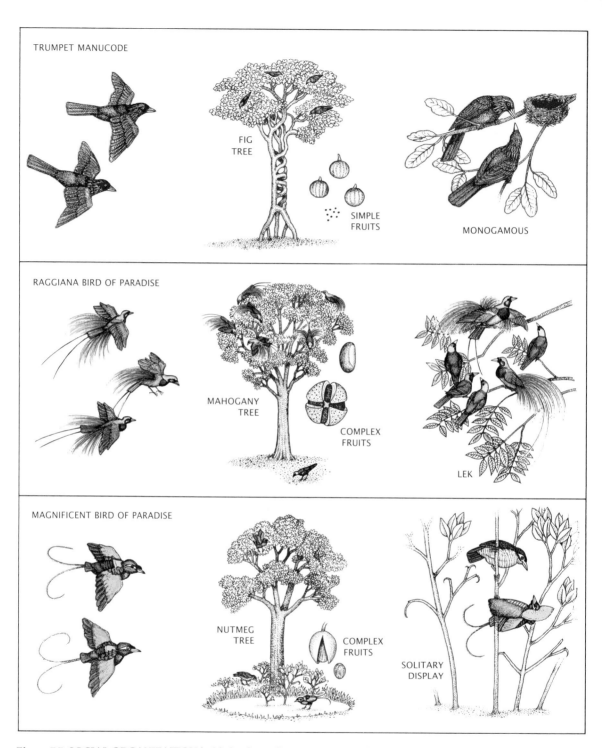

Figure 7.7 SOCIAL ORGANIZATION in birds of paradise is influence by foraging ecology. The trumpet manucode (*top*) is monogamous; males and females cooperate in raising young on a diet of fig pulp. The raggiana (*middle*) is polygynous; males form leks and court actively, but do not help raise offspring. Raggianas feed primarily on complex fruits. The magnificent (*bottom*) similarly is both polygynous and dependent on complex fruits, but males are solitary. Each establishes a cleared display ground in a thicket, where it sings and courts any female that visits.

that bear those fruits are visited predominantly or exclusively by the polygynous birds of paradise. How did such an exclusive association come about? The answer has to do in part with a biogeographic twist of fate.

Monkeys and squirrels (and indeed most placental mammals) have been unable to disperse across a deep-water barrier called Wallace's Line, which separates Australia and New Guinea from the shelf islands of Southeast Asia. In tropical habitats west of the line, monkeys and squirrels harvest and consume virtually all fruits, even well-protected capsular varieties. In many instances the plants derive no benefit from the arrangement: unlike the birds of paradise, the mammals often destroy or consume seeds rather than dispersing them. In New Guinea few mammalian seed predators exist, and so the birds of paradise have little competition for the highly nutritious capsular fruits.

But there is more to the story. Pratt and I have shown that the polygynous Paradisaeidae are among the best avian seed dispersers in the rain forest. The birds eat small numbers of fruits at one time and digest only the seed's nutritive covering, not the seed itself. Moreover, during their daily foraging period the birds actively fly from one tree to another; as they move about, the birds scatter the seeds throughout the forest and do not deposit them in clumps as other fruit feeders often do. Having well-protected complex fruits that are inaccessible to most foragers ensures that the seeds of these plants are handled by effective dispersers that are attracted to the fruit rewards the plants offer.

Compare the foraging strategies of the monogamous trumpet manucode with those of the polygynous magnificent bird of paradise. The trumpet manucode is a specialist: it is not particularly adept at foraging for insects, and so it feeds mostly on figs. Although there are advantages to such a diet—figs are superabundant, easy to harvest and easy to digest—there are also disadvantages: the figs are available at unpredictable times (and so the birds must continually search for them), and they are also poor in nutrients essential to growth and development. As a consequence, the male must help the female to forage for food to feed their nestlings.

The magnificent bird is also a specialist in that it feeds mostly on complex fruits, but the key determinant of its social behavior is the extent to which it supplements its diet with insects. The advantage to feeding on a mixed diet of complex fruits and insects is that both are temporally and spatially predictable and so provide the magnificent with a reliable source of food year round. More important, such a diet is sufficiently high in nutrients that a female can feed her nestlings without the help of a partner. Magnificent males are thus free to spend their time maintaining display sites, which they leave only periodically to obtain a daily ration of food.

From our studies we were able to conclude that the shift from monogamy to polygyny has occurred in those species whose diets include complex fruits in addition to some insects and simple fruits. In species that feed almost exclusively on figs, such as the trumpet manucode, the shift to polygyny has not occurred—probably because both males and females are needed at the nest.

What do such findings reveal about lek formation in the birds of paradise? Why does the raggiana, for example, display in tightly clustered leks, whereas the magnificent and the sicklebill display as solitary males? Jack W. Bradbury of the University of California at San Diego postulates that the evolution of lek behavior is linked to the availability of females, which in turn is determined by foraging behavior and ultimately by diet. He believes that leks form when the females forage far and wide for food and establish overlapping ranges; under these circumstances lekking males have the potential to attract many of the wide-ranging and nonterritorial females.

Bradbury's theory accords well with our data. Females of such well-studied species as the raggiana and Lawes' parotia have large overlapping ranges. A single lek site is likely to be visited by many females, and so the potential for polygyny in these species is high. In other species, such as the magnificent or the sicklebill, the females are not so wide-ranging but forage instead in relatively small, nonoverlapping patches of forest. As a result, not many of them are likely to encounter a display site, and so the potential for polygyny in these species is low.

One can conclude that polygyny and lek formation are promoted by three factors. First, the clumped distribution of fruits creates conditions under which males stationed at a single site are likely to make contact with many females. Second, the predictability and nutrient content of complex fruits make it possible for females to provision their nestlings without the help of a mate and for males to shift to a promiscuous mating strategy. Third, a greater dietary reliance on fruits promotes the expanded ranges of females and the shift of males to lek groups. The raggiana and parotia, for example, feed almost exclusively on complex fruits (rarely on insects) and form leks, whereas the magnificent and the sicklebill frequently feed on insects and are solitary.

S tudying the evolution of mating behavior in the Paradisaeidae from an ecological perspective has led to a new understanding of diet as a determinant of social organization. It is now clear that the birds and the plants on which they feed are interdependent. In addition, my colleagues and I are now aware that only through an understanding of the birds' relation to their rain-forest habitat do the diverse reproductive strategies in this remarkable avian family make sense.

Still, many questions have yet to be answered. For example, will the generalizations we have made about diet and behavior hold true for the birds of paradise that have yet to be studied? On what basis do females discriminate among lekking males, and why do some species have exploded leks, whereas others have tightly clustered ones? And finally, why do males maintain leks when the competition in them is so great? These and other sociobiological riddles will continue to attract biologists for years to come.

Cooperative Breeding
in the Acorn Woodpecker

*The birds share mates and raise their young in groups. Study
of the acorn woodpecker's unusual social system shows how
natural selection yields both cooperation and competition.*

· · ·

Peter B. Stacey and Walter D. Koenig

August, 1984

In the northern temperate regions of the earth almost all species of birds breed in monogamous pairs. The two birds, either singly or together, choose the site for a nest, build the nest, gather food, incubate the eggs, feed the young and guard them. Thus in most species each monogamous pair of birds constitutes a more or less self-contained social unit. A notable exception to this pattern is the acorn woodpecker (*Melanerpes formicivorus*), a common and conspicuous resident of oak and pine-oak woodlands in the southwestern U.S., Mexico and Central America (see Figure 8.1). The social unit of the acorn woodpecker is a territorial group that can include more than a dozen members. In winter the group's main source of food is mast: acorns or other nuts stored in holes drilled in a tree that serves as a storage facility. Such a tree, which is called a granary, is held collectively: the drilling of new holes, the storing of mast and the defending of the tree against intruders are done by all the group members.

Furthermore, in the acorn woodpecker group both mating and the rearing of young are also collective tasks. The sexually mature birds in each group are divided into breeders and nonbreeding "helpers." Among the breeding adults mate-sharing is common. For example, in a group with three breeders of each sex all three males apparently can breed with any female. Thus acorn woodpecker young are truly progeny of the group, and the adult members of the group all contribute to the raising of the young birds. The degree of mate-sharing and the ratio of helpers to breeders depend partly on the climate of the region and partly on the available food resources. As a result there is considerable variation in group structure from one region to another.

The overall mode of breeding of the acorn woodpecker raises difficult questions for the evolutionary biologist. Two central questions are: Why do the helpers remain in the group during the breeding season instead of leaving to reproduce on their own elsewhere? Why do the breeding adults cooperate in raising nestlings that may not be their own offspring? Most students of evolution agree that in general selective pressure operates at the level of the individual organism or at the level of the genes and not at the level of the social group or the species.

Figure 8.1 ACORN WOODPECKERS are highly visible residents of oak and pine-oak woodlands in the southwestern U.S., Mexico and Central America. The birds are about nine inches long. The only visible difference between the male and the female is that the female has a black band separating the red and white areas on the head; here a female is with two males. The acorn woodpecker lives in social groups consisting of as many as 15 members, including several breeders of each sex and additional nonbreeding helpers that are offspring from previous years.

Evolution encourages reproductive strategies that enable an organism to contribute copies of its genes to the next generation. A bird's serving as a nonbreeding helper or raising young that are not its own offspring appears to run counter to evolutionary pressure. Why, then, does such behavior survive in the acorn woodpecker? The answer can illuminate how underlying selective pressures are manifested in social systems that appear to be paradoxical.

Over the past decade we and our colleagues studied the cooperative breeding of the acorn woodpecker in three different U.S. habitats. One of us (Stacey) worked mainly at Water Canyon in the Magdalena Mountains of central New Mexico and also at the Appleton-Battell Research Ranch in southeastern Arizona. The other (Koenig) worked at the Hastings Natural History Reservation in central coastal California. In most areas where acorn woodpeckers have been studied they live in permanent social groups consisting of as many as 15 members. Each group occupies a territory that is vigorously defended from all other acorn woodpeckers. There is generally only one active nest in a group at any one time.

The feature of acorn woodpecker behavior that first attracted the attention of ornithologists was not the fact that the birds live in groups but their habit of storing large quantities of acorns, piñon nuts and other mast in a granary tree (see Figure 8.2). In the fall the group members harvest the mast and wedge it into holes drilled in bark or in a dead tree. Some granary trees have more than 30,000 holes. Storage facilities with this kind of capacity enable the group to protect large quantities of food from the effects of weather and from other animals.

When ornithologists first noted how the acorn woodpecker stores mast, some of them suggested the birds could be employing the mast to farm insects. It was argued that as the mast decomposed over the winter it would be invaded by insect larvae, and the woodpeckers would have the insects as a ready food source. Closer field observations showed, however, that the birds regularly move the mast from hole to hole to keep it from rotting or falling out of the tree as the nuts dry and shrink. The mast itself serves the woodpeckers as food, and mast that has been invaded by insects is rejected.

Any bird that is a member of the group can feed on the mast at any time. Over the winter the stored mast is a highly significant resource for the survival of the group. In the spring and summer the woodpeckers eat insects, tree sap, leaf catkins and flower nectar. In most temperate habitats, however, such foods are either scarce or completely unavailable in winter. An acorn woodpecker group that exhausts its stored mast in the winter months is often forced to abandon its territory.

Many granary trees are easily accessible to the student of the acorn woodpecker, and with patience the stored acorns or nuts can be counted. In this way the energetic value of the stores and of the area's total woodpecker population can be estimated. The estimates show that stored mast is critical not only for survival but also for reproduction. For example, in a 10-year period at the Hastings reservation in California, of the groups that possessed stored mast in the spring, 83 percent successfully raised young birds to the stage of fledging, or leaving the nest. Of the groups that had exhausted their stores only 20 percent successfully raised young to fledging. In years when the acorn crop is particularly large groups sometimes breed in the fall as well as the spring, a phenomenon rare in temperate climates.

The effect of mast storage was also observed at Water Canyon in New Mexico. Over a nine-year period groups that had mast in the spring fledged an average of 2.7 young per year and groups that had no mast fledged an average of 1.3 young. The results are particularly striking because the mast is rarely fed to the nestlings. Instead it functions as a food reserve that enables the adults to search for insects, which are then fed to the young birds.

By noting at regular intervals the amount of mast stored by different groups and correlating the counts with observations of the groups it was shown that mast storage has a strong influence on the structure of the acorn woodpecker population in a particular area. In years when the quantity of stored reserves is comparatively high the fraction of adults that survive the winter is also high. As a result the total woodpecker population of the area is large and more groups are able to remain in their territories until spring. Since most territories are continuously occupied, few young birds can leave to colonize territories of their own. Therefore a relatively large fraction of each group is made up of nonbreeding helpers (see Figure 8.3).

The amount of mast a woodpecker group can store is determined partly by the yield of acorns or nuts from the trees in its territory. It is also influenced by the number of storage holes the group has drilled in the granary tree. Since the amount of stored mast has such a strong influence on the structure and continuity of the group, it would appear that group members should spend much of their time drilling new storage holes.

One of us (Stacey) found, however, that the birds drill storage holes only when they have a supply of mast. When the food stores are exhausted, no drilling is done. The reason is that in the absence of accumulated mast the members of the group spend much of their time looking for other food. If a store of mast is available, less time is spent in foraging. Furthermore, some members of the group must be near the granary to guard the stores, and it is the guards that drill the new holes; if there are no stores to guard, the guards too are off looking for food.

The construction of storage facilities by the acorn woodpecker group is an example of a biological positive-feedback system. Groups with many storage holes can accumulate enough mast to enable them to drill new holes over the winter. The enlarged capacity makes it possible for the group to store more nuts the following fall. Over a long period the number of storage holes could be limited only by the fact that the granaries do not last forever.

The granaries are often constructed in trees that are dead or partly so. Even the sturdiest storage trees, which are oaks, typically last for only a few decades from the time the woodpecker group begins to utilize them. Eventually the granary tree rots and falls. The fall of the tree leads to a reduction in the group's size, in its reproductive success and in the probability that it will remain resident in its territory throughout the year. In some instances the fall of a large granary tree can even lead to the dissolution of the group.

Research in recent years has shown there is a considerable diversity in the reproductive organization of species of birds that breed cooperatively. In some species, including the green wood-hoopoe of Africa and the Florida scrub jay, the birds form groups made up of a single breeding pair and nonbreeding helpers; the helpers are generally offspring from previous years. Other species, including the Tasmanian native hen and the Galápagos hawk, show cooperative polygamy, or mate-sharing. Mate-sharing entails the presence of several male or female breeders; in both the Tasmanian and the Galápagos species the multiple breeders are always males. In such species it is thought the brood can receive genes from all the male breeders. Even among such unusual species the acorn woodpecker stands out because the woodpecker groups have both nonbreeding helpers and mate-sharing. Moreover, the mate-sharing often involves males and females simultaneously.

In the acorn woodpecker, courtship behavior between male-female pairs is absent. In addition the entire group tends a single nest. As a result determining the reproductive roles in a cooperative group is not an easy task. Copulation is rarely observed, and so it is particularly difficult to know the reproductive status of male birds. It had long been suspected that more than one male in a group can breed, but proof was lacking. It has recently been obtained by genetic analysis of enzymes in the blood of members of several woodpecker groups; this work was done in collaboration with Nancy E. Joste and J. David Ligon of the University of New Mexico and Ronald L. Mumme and Robert M. Zink of the University of California at Berkeley (see Figure 8.4).

The reproductive status of females is easier to determine because egg laying can be observed if the group is watched carefully during the breeding season, which is generally in late spring and early summer. At Water Canyon woodpecker groups with more than one breeding female are rare. At the Hastings reservation, however, 20 percent of the groups include two or more females that lay eggs in the same nest. In both areas the groups are supplemented by males and females that do not breed.

With both the Water Canyon and the Hastings reservation woodpecker populations one of the first lines of work was to unravel the reproductive relations in the group and find out what determines the reproductive status of individual birds. It developed that there are intriguing differences in structure between groups in one of the areas and those in the other. In both areas helpers are group offspring from previous years. At the Hastings reservation, however, the males that share mates are frequently brothers and less frequently father and son; the females that share mates can be sisters or mother and daughter. At Water Canyon the birds that share a mate can be genetically related or unrelated.

To understand how such differences could arise the history of a typical group must be considered. A group can be founded by a male-female pair or by the joining of two "coalitions" from different groups. One coalition consists of from one male to four males, usually siblings, drawn from an established group. The other consists of from one female to three females, also usually siblings, drawn from another group.

Once the coalitions have merged, all the mature adults can breed as a unit. The young birds that are subsequently fledged remain in the group at least until their first spring and sometimes longer. With one notable exception, which will be discussed below, the group offspring do not breed as long as they remain in their natal group. The group varies in size as young birds are fledged and as they and other birds leave. The unit maintains its continuity as long as at least one of the original breeders of each sex is present. As the original breeders die or

Figure 8.2 GRANARY TREE is utilized by the acorn woodpecker group for the storage of mast: acorns and other hard-shell nuts. The tree is an oak at the Hastings Natural History Reservation on the coast of central California. The holes are drilled by successive generations of the members of a single acorn woodpecker group; some granary trees are perforated with as many as 30,000 holes. The nuts are inserted in the holes in the fall as the group attempts to accumulate enough food stores to last through the winter. The granary tree is a central feature of the ecology and social structure of the woodpecker group.

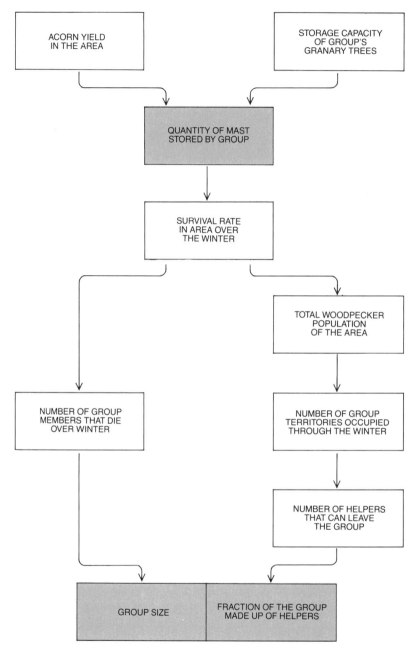

Figure 8.3 ACORN WOODPECKER GROUP STRUC-TURE is strongly influenced by the quantity of mast the group can store in the fall, which is itself determined by the yield from the trees in the group's territory and the storage capacity of the granaries held by the group. The quantity of mast affects how many birds survive the winter, which influences the total local population and the number of desirable territories that remain occupied all year. If the population is large with many territories occupied, few young can leave their parental groups to breed on their own. As a result the groups are large and a substantial fraction of each group consists of nonbreeding helpers. The quantity of stored mast also influences the number of young raised by the group.

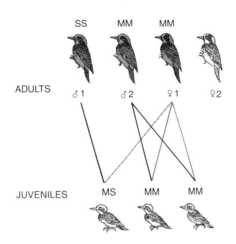

Figure 8.4 FIRST DEMONSTRATION of multiple paternity in an avian communal breeder was obtained for the acorn woodpecker by the authors and their co-workers. The paternity relations in a group at Water Canyon, N. Mex., are shown. The letters *S* and *M* stand for slow band and medium band. Each letter designates an allele, or variant of a gene, for a blood enzyme. Each pair of alleles was identified by gel electrophoresis, in which enzymes can be separated according to their mobility in a gel. Female No. 2 was a nonbreeding helper.

leave, however, there comes a time when all the founders of one sex or the other have gone.

The disappearance of all the original male breeders or all the original female breeders creates what is known as a reproductive vacancy, which is of fundamental significance in the history of the group (see Figure 8.5). The reproductive vacancy is not filled by helpers from the group; a daughter does not replace her mother as a breeder, nor does a son replace his father. Instead the gap is filled by mature birds from outside the group.

Contrary to what was thought only a few years ago, young acorn woodpeckers do not remain in their natal territory waiting to inherit it from parents of the same sex. Once the reproductive vacancy has been filled, however, group offspring of the sex opposite that of the new breeder can mate. For example, if a female breeder in a group has been replaced, the group offspring that are male can breed in spite of the presence of the original male breeders; the converse is true for the females.

Explaining the features of reproduction among acorn woodpeckers is harder than describing how the group forms. The sharing of mates does provide one explanation for why a group of birds will cooperate in feeding the young. The reason is that in some groups all the cooperating adults are actually or potentially the parents of the young birds. The hypothesis does not, however, explain the role of the helpers or the willingness of the breeders to share mates.

Indeed, it might be supposed that mate-sharing runs counter to the reproductive interest of each breeding adult. The number of young that can be cared for by the group is limited. Hence when the breeder shares a mate, there is a reduction in the fraction of the group young that are the offspring of that breeder. In most species of birds that cooperate in raising offspring there is no mate-sharing; the dominant male and the dominant female exclude all other adults from breeding and only a single pair reproduces.

Recent work by one of us (Stacey) suggests, however, that under certain circumstances mate-sharing is to the advantage of the individual breeder. In some environments cooperation makes a significant contribution to the individual's reproductive success and even to its long-term survival. The reduction of group size would put the individual at a disadvantage. If subordinate birds that were excluded from breeding were likely to leave the group, the dominant birds could enhance their own reproductive success by sharing the breeding.

Data collected at Water Canyon and at the Hastings reservation support the idea that such factors are operating in acorn woodpecker groups. At Water Canyon the number of young that are fledged per breeding male is higher in groups consisting of two males and one female than it is in breeding pairs. In the groups of three there were 1.16 young per adult male on the average; in the pairs there were .92 young per male. Thus when a male shares paternity with another male in a threesome, he will fledge more of his own young than he would if he evicted the second male and bred by himself.

At the Hastings reservation the reproductive success of the individual decreases as the group expands but the fraction of birds that survive the year

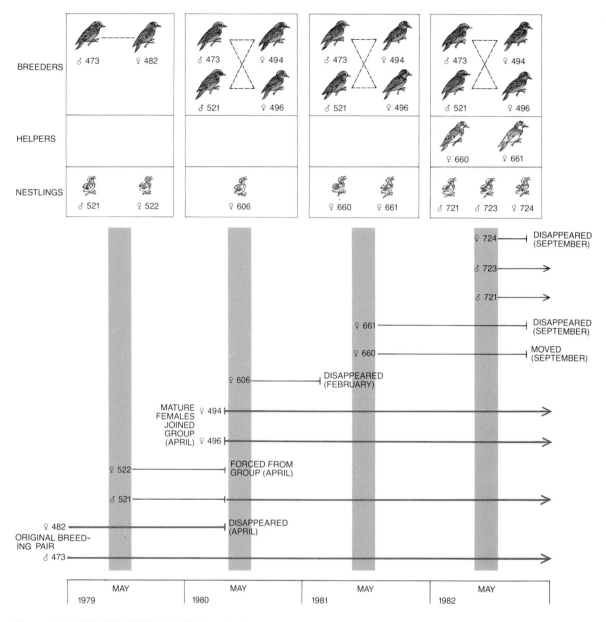

Figure 8.5 REPRODUCTIVE VACANCY, the disappearance of all the original male (or female) breeders, is a major event in an acorn woodpecker group. In this history of the Hastings Plaque group (1979–1982), color indicates breeding adults. In May, 1979 the original pair (male #473, female #482) raised two young (male #521, female #522), which could not breed while their parents remained in the group. However, the mother vanished before the 1980 breeding season and two sisters from a nearby group (#494 and #496) won the ensuing power struggle, evicting daughter #522 in the process. With his mother gone, male #521 could breed with the new females, as could his father.

increases. A male or a female that is the only breeder of its sex in a group has an annual survival rate of 70 percent. Among females sharing a mate the survival rate is 79 percent and among males sharing a mate the rate is 86 percent. Although the rates were lower at Water Canyon, the pattern was similar: survival was 47 percent for birds breeding in pairs and 65 percent for those breeding in large groups. When the data for reproductive success and survival are combined, it becomes clear that in both the Water Canyon and the Hastings reservation populations a bird that begins its reproductive career sharing a group with other individuals of the same sex has an advantage over one that breeds in a pair

Although at first mate-sharing and cooperative breeding appear to contradict the principle that natural selection operates on the individual organism, close examination shows that cooperative activity benefits the individual breeder. Not all the activities in the acorn woodpecker group, however, are harmonious. Selective pressure has also encouraged intense competition. Two forms of reproductive competition are notable: egg-tossing and the killing of young birds.

Egg-tossing in acorn woodpeckers was first observed by Ron Mumme at the Hastings reservation (see Figure 8.6). As we have mentioned, many of the breeding females in a group are closely related genetically. A typical egg-tossing episode could involve two sisters. In the breeding season the sisters lay their eggs in periods that nearly overlap but are not precisely synchronized. Each female lays some three to five eggs at the rate of one egg per day. The eggs are laid in the same nest cavity and the sisters share the incubation and the feeding of the young.

The first egg to be laid triggers a competition between the females. When one of the females lays the first egg, it is almost invariably removed from the nest by the sister. The sister takes the egg to a nearby tree and puts it in a depression in the bark of a limb. Then the egg is pecked open and the contents are eaten by several group members, often including the female that laid the egg.

When the female that laid the first egg lays another on the following day, the egg-tossing is repeated. Observations have been made of as many as five eggs being consecutively tossed from a nest. The egg-tossing stops only when the second female begins to lay. At that point the bird apparently cannot distinguish her eggs from her sister's and therefore refrains from further destruction.

The adaptive value of egg-tossing is clear. The female that destroys her sister's eggs gains a genetic advantage in the clutch of eggs incubated by the group. If the egg-tossing had not taken place, each sister would have contributed an equal number of genes to the next generation. After the eggs are tossed the bird that did the tossing contributes a majority of the genes passed to the offspring. A pair of sisters may live together in a group for several years, however, and the female that has her eggs tossed one year may toss her sister's eggs the next year. The reproductive success of the sisters over their life span could be approximately equal.

An even more extreme form of reproductive competition is the killing of young birds by recent immigrants to the group. Among animals that live in groups the killing of young by recent arrivals had been noted only in primates and mammalian carnivores such as lions. It was first observed in the New Mexico acorn woodpecker population in 1981.

The killing of young birds generally takes place in the spring or early summer when a woodpecker joins a group with an active nest. A precondition for the killing appears to be that the immigrant is the only mature bird of its sex in the group, either because the previous breeders of its sex have left or because the immigrant has driven them away. Once the new bird is accepted by the group members of the opposite sex, it immediately begins to visit the nest and eventually destroys the eggs or kills the nestlings. The bodies of the young birds are removed from the nest and can be eaten by all the group members.

There is controversy among evolutionary biologists over whether the killing of young in a cooperative group is always an adaptively advantageous move. In the acorn woodpecker group, however, the immigrant clearly benefits. Since there is only one nest, a bird that joins the group after the eggs are laid cannot contribute its genes. If enough time remains for the group to breed again, an immigrant that kills the young forces the group to breed and thereby ensures that its genes are passed on.

Acorn woodpecker groups are complex units in which members cooperate and compete to maximize their genetic contributions to the next generation. Why the intricate group structure evolved is not known with any precision. Some of the environmental factors that influence the group, however, are fairly well understood, including the factors that determine the status of the nonbreeding helpers.

Most group offspring remain in the natal group as

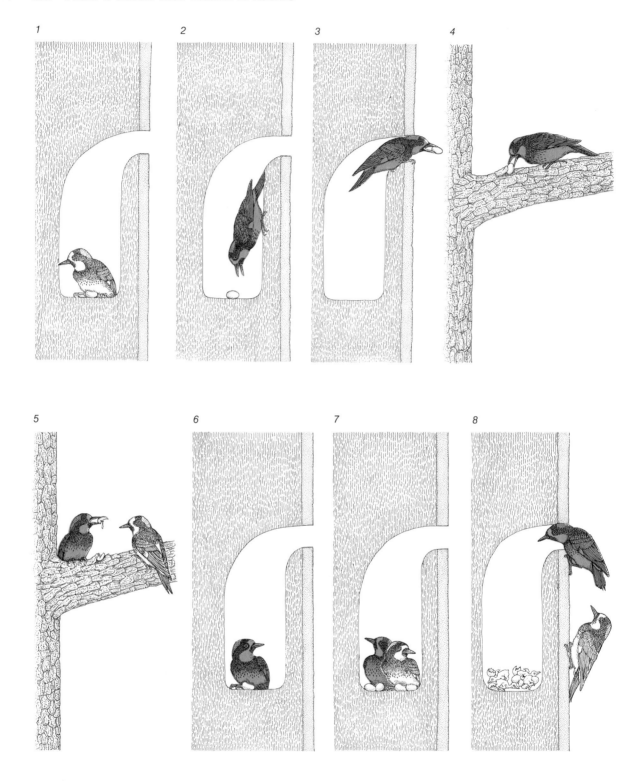

helpers because they are ecologically constrained from leaving; if the opportunity to leave arises, the young birds generally disperse to form a new group. The constraint on the helpers appears to be habitat saturation, which means simply that there are no empty territories suitable for occupation by new groups. Habitat saturation does not imply that every square inch is occupied by an established acorn woodpecker group. In any given area, however, the optimum habitats are continuously occupied. Furthermore, in most regions where acorn woodpeckers live there is a shortage of marginal habitats to be taken over by a new group made up of birds that had not previously been breeders. Since group offspring cannot breed independently, they might benefit from remaining in their parents' territory and helping to raise the offspring of their genetic relatives.

Several lines of evidence support the conclusion that habitat saturation has a significant role in keeping young acorn woodpeckers in their natal group. One striking item of evidence is the dramatic power struggle that takes place over reproductive vacancies. Frequently sets of siblings from groups several miles apart converge on a territory where there is a vacancy; the birds fight for the right to remain. Such struggles, which can engage as many as 50 birds and can last for several days, consists of uninterrupted chasing and territorial displays (see Figure 8.7).

The winners in the struggle fill the vacancy and the losers return to their parental group. The ecological significance of these chaotic struggles is that nonbreeding helpers are willing to fight intensely for the opportunity to leave the natal group and breed on their own. Such struggles are logical only if good habitats are at such a premium that the young are constrained from leaving the parental group.

Figure 8.6 EGG-TOSSING is a form of genetic competition between acorn woodpecker females, even sisters. Each lays one egg per day, but their laying periods are not exactly synchronized. Soon after one female (*white*) lays the first egg (1) her sister (*color*) removes it (2, 3), places it in a depression nearby (4) and shares its contents with other group members, often including the egg's mother (5). Such egg destruction is repeated until the second female begins laying in the same cavity (6) and tossing stops (7). Group members incubate until hatching is complete (8). For the season, the tossing female ends up with more eggs in the clutch than her sister, but the roles may change next year. Thus it is possible that both make similar genetic contributions in the long run.

The lack of marginal territories suitable for colonization by newly formed groups is apparently due to the fact that the woodpecker relies heavily on the granary tree. The granary is the result of work done by generations of woodpeckers, with each bird drilling a few holes per year. A threshhold of from several hundred to a thousand storage holes must be reached before a territory can be permanently occupied. When enough holes have been drilled, the territory can be occupied continuously until the granary tree falls. Individual birds cannot create a new granary in a short period, and therefore they cannot disperse into a new area unless it already has a granary tree. Marginal habitats are scarce because all mature granaries are in occupied territories.

Several other workers, including Jerram Brown of the State University of New York at Albany and Stephen T. Emlen of Cornell University, have suggested that ecological constraints are one of the main causes of cooperative breeding in birds. The study of the acorn woodpecker in several different regions has made it possible to carry out the first quantitative test of that hypothesis. The level of habitat saturation can be measured according to the number of territories that become vacant over the winter, which is the time young birds attempt to leave the natal group.

At the Hastings reservation most groups can store enough mast to last through the winter and hence most territories are occupied continuously. In Water Canyon, however, the yield of acorns or other nuts is more variable; many groups exhaust the stored food and abandon their territory. The vacant territory is then available for colonization. As is predicted by the habitat-saturation hypothesis, fewer young birds remain in their original group for a year or more at Water Canyon than remain at the Hastings reservation.

Additional evidence that habitat saturation and the construction of mast-storage facilities are among the causes of cooperative breeding in the acorn woodpecker group comes from a study done in southeastern Arizona by one of us (Stacey) in collaboration with Carl E. Bock of the University of Colorado at Boulder. In this area the annual acorn yield is sparse and variable; there are few years when the woodpeckers can store enough mast to support the group over the winter. As a result groups do not construct granaries.

Instead acorns are gathered and stored in natural cracks and holes in the bark of trees. The capacity of the storage system is small, and the mast is gener-

Figure 8.7 WAKA DISPLAY is a gesture made by an acorn woodpecker to other members of its group while guarding the granary and also in the power struggle over a reproductive vacancy within a group; the raising of the wings constitutes a "greeting" or "recognition" display among group members. The power struggle can involve as many as 50 birds that engage in several days of continuous chasing and display. The winners of the struggle remain to breed in the empty territory; the losers return to the group where they were born and resume the function of helper.

ally exhausted in the fall soon after the oaks stop producing acorns. When the food store is used up, the birds leave their territory and migrate to Mexico for the winter. In the spring many of the same birds return and establish new breeding territories. Because of the annual migration most of the breeding habitat is unoccupied in the winter and therefore much territory is available for colonization in the spring.

The absence of permanent territories has a profound effect on the birds' reproductive behavior. In contrast to the young birds in California (Hastings reservation) and New Mexico (Water Canyon), young birds in southeastern Arizona leave their natal territory in the fall and do not migrate to Mexico as part of a family group. In the spring the adults and yearlings return to Arizona individually and often breed in a new territory with a new partner every year. Even more striking is the fact that in Arizona the migratory acorn woodpeckers do not breed cooperatively. There are no nonbreeding helpers and mating is in isolated pairs.

From the work done on the acorn woodpecker in the past decade have come new perceptions of the ecology and social organization of cooperative breeders. Such studies have given a detailed picture of the birds' reproductive behavior and their mating system. Above all, the work has illuminated how intricate the relations are among members of an avian social group.

Biological evolution favors individuals that consistently maximize their genetic contribution to the next generation relative to the contribution of other breeders. Yet the strategies by which the genetic contribution is increased are not simple. The strategies adopted by the acorn woodpecker range from the straightforward genetic competition entailed in destroying a sister's eggs or killing the young of other group members to the apparently self-denying behavior of mate-sharing or acting as a nonbreeding helper. Further work on this unusual bird will undoubtedly reveal other unexpected means whereby a woodpecker in a social group can increase its reproductive advantage and hence its evolutionary fitness.

The Cooperative Breeding Behavior of the Green Woodhoopoe

Among these East African birds one pair in each flock breed while the nonbreeding adults assist in the raising of the breeders' chicks. How did this pattern of seemingly altruistic animal behavior evolve?

. . .

J. David Ligon and Sandra H. Ligon
July, 1982

Over the past few years investigators of animal behavior have devoted much attention to the evolution of complex social organizations. One aspect of social behavior is cooperation: exchange between individuals of some kind of resource or assistance. Cooperation is common in human societies and in general is mutually beneficial. The giving of assistance can, however, be unequal and even one-sided (as in life-sacrificing human heroism). The commonest instances of unequal exchange among the lower animals are found in the social insects. In these animal societies workers are sterile and spend their entire adult lives engaged in activities that benefit the reproduction of the queen. Often they give their lives to defend the nest. Such behavior seems to go beyond the bounds of cooperation and to meet the biologist's definition of altruism: "behavior that decreases or potentially decreases the lifetime reproductive output of the altruist to the benefit of another individual or individuals." This definition does not, of course, imply any conscious reasoning or forethought.

Although the social insects provide the most striking examples of apparent altruism, complex levels of cooperation, including apparently unequal aid giving, are also found among higher animals other than man. A vivid case in point is a species of insect-eating bird of Africa: the green woodhoopoe (*Phoeniculus purpureus*). The birds' social unit, the flock, may have as many as 16 members but only one breeding pair. The other sexually mature but nonbreeding flock members serve both as "nest helpers," sharing the burden of bringing food to the incubating female and later to the nestlings, and as "guards," defending the nestlings and later the fledglings against predators and participating throughout the year in such flock activities as protecting the home territory.

More than 100 species of birds share with the green woodhoopoe this pattern of cooperative breeding, but no single set of factors can explain why one species displays the behavior pattern and an apparently similar species does not. To understand the selective pressures that favor the evolution of cooperative breeding one must take into account the influence of the environment on a given social system. Among the environmental factors are the climate, the distribution of food both in time (with the seasons) and in space (within the utilized area), the availability and quality of roost sites and nest sites and the kinds and numbers of predators. Another important component of natural selection is what may be called the social environment: the effect on the individual of interaction with other members of its own species.

We began our work with the woodhoopoes in 1975. The birds were well suited for study. The flocks we observed inhabit acacia woodland near Lake Naivasha in Kenya (see Figure 9.1), where they roost at night inside cavities in the trunks of the acacias, some of them naturally formed and some unoccupied woodpecker holes. This makes individuals easy to catch and mark; one simply plugs the mouth of the cavity after dark and then lets the bird emerge into a transparent bag the next morning. Both the openness of the woodland and the sparseness of the acacia foliage made observing the birds fairly easy; so did their generally calm disregard of human beings and the birds' comparatively large size. (The males are about 36 centimeters long and the females about 30.)

During our first period in the field, from July, 1975, through May, 1976, we color-banded 151 birds, a total that included nearly all the members of 25 flocks. We subsequently banded an additional 218 birds, most of them offspring of the first flocks we had banded. As a result our field notes include the sex, the age and the flock affiliation of the birds and also the parentage of nearly all of the second group.

The birds in our study area were highly territorial. It was unusual when a suitable habitat, for example a patch of woods that included roosting cavities, was not "owned" by one flock or another. Each territory was vigorously defended both against intrusion by birds from neighboring flocks and against individual "outsider" woodhoopoes that occasionally wandered into the area (see Figure 9.2). This pattern of behavior made the establishment of new territories rare. The groups varied in size, some consisting only of the breeding pair. Generally, however, the flocks were larger, and the birds that "helped" the breeding pair were usually, although not invariably, the siblings of one breeder or the offspring of one or both birds in the breeding pair.

The woodhoopoes in our study area did not on the average live long. Each year between 30 and 40 percent of the population died. The distribution of the death rate in the population was uneven. Breeding males were the most frequent casualties, their deaths outnumbering those of the other three categories: breeding females, female helpers and male helpers (see Figure 9.3). Most of the deaths were caused by predation. The primary predators appeared to be genets (catlike members of the mongoose family) and driver ants (the African counterpart of the New World army ants). Both predators raid the woodhoopoes' roosts at night. On several occasions we found the remains of dead birds still in the roost holes or on the ground below the nest sites.

The environmental factors in addition to predation also have a great effect on the woodhoopoe population. The first is the timing and amount of rainfall. The rains in this region of Kenya are highly variable. The "normal" annual pattern includes a dry season that extends from December through February, followed by the "long rains," which begin in March or April. Over the seven years of our study the woodhoopoes began their breeding activity in May or June, a few weeks after the onset of the long rains.

Abnormal rainfall can sharply limit the woodhoopoes' main food supply: the larvae of some 2,000 species of moths that inhabit the woodland. If the dry season is truly dry, the pupating moths thrive in their buried cocoons. As the rains begin the moths emerge to breed and lay the eggs that give rise to the next generation. If, however, there is rain in the dry season, the pupae may rot or be destroyed by fungi and other microorganisms. With fewer pupae to become moths and lay eggs there are fewer caterpillars.

We witnessed such an unfavorable state of affairs in 1979. That year nearly 30 centimeters of rain fell in January and February, compared with a 34-year average of less than six centimeters. The woodhoopoes started nesting in June, but only two of the 11 flocks in our study area managed to raise any young. Exactly the opposite happened in 1981. Only 2.1 centimeters of rain fell in January and February, and the moth pupae thrived. The long rains of March through May were unusually heavy: 36 centimeters. All 19 of the woodhoopoe flocks in our area that year, including four pairs with no helpers, nested in May and June. Reproductive success was unusually high, with all but two groups producing young. It would have been higher still if predators had not eaten two broods of nestlings. This large annual variation in food supply (and re-

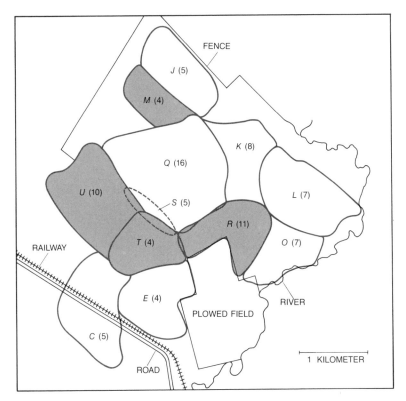

FENCE

J (5)

M (4)

K (8)

Q (16)

U (10)

S (5)

L (7)

R (11)

T (4)

O (7)

RAILWAY

E (4)

RIVER

C (5)

PLOWED FIELD

ROAD

1 KILOMETER

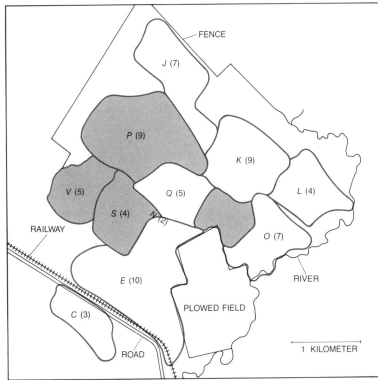

FENCE

J (7)

P (9)

K (9)

V (5)

Q (5)

L (4)

S (4)

N (2)

RAILWAY

O (7)

E (10)

RIVER

C (3)

PLOWED FIELD

ROAD

1 KILOMETER

Figure 9.1 OPEN ACACIA WOODLAND on a large farm in the Rift Valley of Kenya was the area selected for the woodhoopoe study. In the late 1975 (*top*) the 2,000-acre site was inhabited by 11 established flocks and a 12th flock in the process of establishment (*broken boundary line to the left of center*). The numerals indicate the number of birds in that flock at the end of the year. The total was 86. The four territories shown in color no longer existed by late 1981 (*bottom*), but four new territories (*gray*) had been established. The number of birds then was 65.

Figure 9.2 TERRITORIAL DISPUTE brings three birds from one flock to a confrontation with four from a neighboring flock. The dispute is chiefly vocal. One bird in each group, however, stands with wings loosely held.

productive success) represents one critical environmental variable for the woodhoopoe.

Another variable, perhaps equally critical, consists of the number of roost holes available to the woodhoopoes. The birds compete for the cavities with birds of other species and also with honeybees and small mammals. The presence or absence of roost holes in a particular woodland area determines both the distribution of woodhoopoe flocks and the reproductive success of the flocks over the years. Suitable roosts are often so scarce that most or all members of each sex roost together. Such "dormitory" behavior may be a by-product of sexual dimorphism: the difference in size between the males and the females. The females, being smaller,

can enter cavities the larger males cannot. In territories where suitable roost holes were rare we have found as many as eight females roosting together in a single cavity.

This segregation by sex has an important territorial implication. A nocturnal predator can be chance eliminate most or all of the males or females in a flock in one night. Such an event leaves the flock's territory open to colonization by birds of the sex that has been eliminated; the colonizers may be either solitary individuals or members of flocks in adjacent territories.

Our years of observation yielded numerous examples of the adaptive and evolutionary sig-

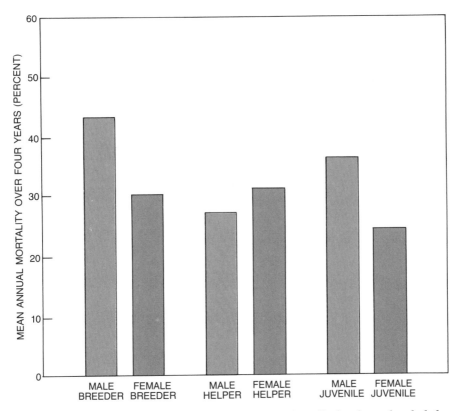

Figure 9.3 HIGH MORTALITY RATE among woodhoopoes over a four-year period (January, 1976, through December, 1979) was not evenly distributed either by sex or by social status. Breeding males and juvenile males led in fatalities; juvenile females and male helpers trailed. The rates for breeding females and female helpers were much alike and fell between those of the others.

nificance of the behavior of the helper woodhoopoes. For example, consider the birds' high mortality rate. For a newly fledged bird to leave its parents' territory and search for a territorial vacancy elsewhere without a known roost for shelter at night is quite risky. It is also indirectly risky for the young bird's parents: in an evolutionary sense producing mature offspring is their only reason for existence. For the young bird to remain in its parents' territory in the role of a nonbreeding helper is far less risky. Since helpers are usually the offspring of one or both of the members of the flock's breeding pair, their remaining with the flock for an extended period is also indirectly beneficial in an evolutionary sense to one or both of the parents. Indeed, we view this pattern of continuing affiliation with the flock as constituting a form of extended parental care.

An observed fact supporting this interpretation of continuing affiliation by the helper birds is that there is no positive correlation between the number of helpers in a flock and the number of offspring produced by the breeding pair in the course of a season. This means the protection that continuing affiliation with the flock gives the helpers does not necessarily increase the reproductive output of the breeding pair and may sometimes even decrease it.

A further aspect of high mortality is the likelihood that one or another of the breeding pair in a flock will die and be replaced within a short span of time. As an example, in 1981, a year of exceptionally high breeding success, the 12 flocks in our study area included only 25 helpers. Of this total five helpers (20 percent) had the same parents as the nestlings the helpers were feeding. The degree of relatedness between the other 20 helpers and the nestlings they attended ranged from zero (three birds known not to share close relatives) to 37.5

percent (six birds with one parent in common with the nestlings and the other parents siblings). In other words, high adult mortality leads to lower levels of relatedness between helpers and nestlings than kinship theory might predict (see Figure 9.4).

If one asks whether helpers have the option of being able to colonize new territory, the answer is no, at least most of the time. For example, the number of woodhoopoes in our main study area fluctuated widely between August, 1975, and August, 1981, ranging from a low of 46 birds to a high of 94 (see Figure 9.5). The number of territories containing a breeding pair at the onset of each breeding season, however, fluctuated far less, ranging from a low of 12 territories to a high of 14. Between 1975 and 1981 four flocks ceased to exist; their former

territories remained empty. In the same period four new territories were developed.

Considering the large variation in the number of individuals over the seven-year period, the relative stability in the number of breeders and territories suggests that the opportunity to colonize a new territory is rare. A pattern of wandering—in effect searching out a new territory—with its attendant dangers is evidently uncommon because it is usually disadvantageous. These considerations help to explain why young woodhoopoes stay with their parents' flock. They do not, however, account for the birds' helping behavior.

That is not to say emigration is always disadvantageous (see Figure 9.6). The adult woodhoo-

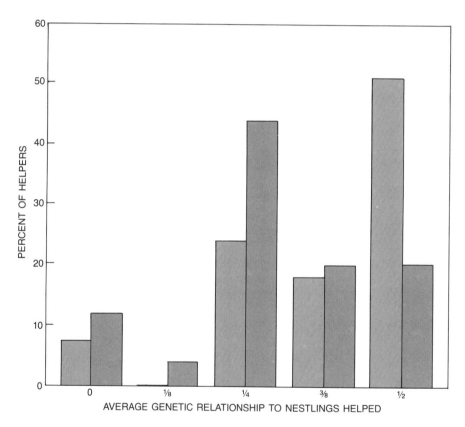

Figure 9.4 TIES OF KINSHIP within individual flocks were increasingly diluted over time. This graph compares the relationship between helpers and the nestlings they cared for in 1978 (*color*) and 1981 (*gray*). In the earlier year 14 flocks included a total of 55 helpers; in the latter year nine flocks included a total of 25 helpers. More than 50 percent of the helpers in 1978 shared common parentage with the nestlings. By 1981 only 20 percent of the helpers did so. Fewer than 25 percent of the helpers in 1978 were the equivalent of half-siblings of the nestlings and none were the equivalent of first cousins. By 1981 these relationships were close to 45 and 5 percent.

poes' high mortality rate means that neighboring territories have frequent breeding vacancies. As a result the commonest way for a mature nonbreeder of either sex to attain breeding status is to emigrate to a territory where a vacancy has been created by the death of a breeder of the same sex (and of mature nonbreeders of that sex, if any).

Emigrants seldom made these shifts alone. We observed that a successful move into a strange territory usually involved a team of two or more birds of the same sex. Such teams were generally composed of flock mates but they never included nest mates; antagonism between former nest mates is strong and they do not emigrate together. As a result the emigrant teams consisted of an older nonbreeder (we labeled them alpha birds) and a younger one or two (beta and gamma birds). A subtle dominance hierarchy, positively correlated with greater age, was apparent. We found that the alpha emigrant became the breeding replacement in the adopting flock. If the alpha emigrant then died, the beta emigrant inherited the alpha's breeding status (see Figure 9.7).

The observed facts of team emigration, with the dominant older partner a helper that had been involved in rearing the subordinate younger partner or partners, gave us an insight into the evolution of helper behavior. For a nonbreeder to achieve breeder status by emigration—one of the two roads to survival in the genetic sense—it was first necessary for it to acquire one or more subordinate allies. Could this be, at least in part, why helpers help? Nothing in this evolutionary process need involve the exercise of reasoning: the genes of unhelpful and therefore unallied nonbreeders are at risk of being eliminated from the gene pool of the species within a single generation.

How is the bond established between the helper and the nestling it helps? We found that helpers not only brought the nestlings food but also stole caterpillars from other helpers in order to deliver them to the nestlings they tended. Helpers interacted with nestlings in other ways (see Figure 9.8). For example, they groomed them. They also perched nearby and vocalized with their attention directed to the nestlings. Neither these activities nor the fetching of food seems to represent a costly investment in time and energy compared with the high genetic return that may result: achievement of breeding status.

The fact that older nonbreeders may benefit in this way still leaves a key question unanswered: Why do younger birds accompany their elders when the opportunity to emigrate arises? Not only do the betas help the alphas to become established in the new territory but also they take up the burdens of a helper when the offspring of the alphas are hatched. The answer again seems to lie in the pattern of woodhoopoe mortality. Without implying any reasoned behavior, an anthropomorphic description of the young birds' actions would not be that they were "paying back" the older birds for their earlier help but rather that they were "playing the odds" in expectation, as it were, of their own genetic reward.

For example, male alphas, it will be remembered, have a significantly higher mortality rate than male betas. As a result subordinate males will frequently outlive their dominant allies and then themselves inherit breeding status. We observed that among groups of two to four emigrant males, and among newly formed flocks composed of only two adult males and one or two adult females, in 10 out of 13 instances the alpha male died before the younger males. We also found that over their lifetime once-subordinate emigrant males left as many surviving offspring as the originally dominant males. This observation bears crucially on the evolutionary concept known as individual fitness. With regard to the beta birds' behavior, as with regard to the alphas', there is no reason to suppose that when they emigrate with older flock mates they are in any way acting contrary to their own long-term reproductive interests.

In this connection we also observed that some female helpers, unlike male helpers, voluntarily left their original territory unaccompanied by subordinates and wandered over a large area apparently searching for a territorial vacancy. This behavior may arise from the fact that the mortality rate for female breeders is significantly lower than that for male breeders. Therefore the likelihood of a female helper's achieving breeder status at an early age in the home flock is remote, and seeking breeder status somewhere else, although it is risky for a solitary bird, may be genetically rewarded.

Since any individual bird's future is uncertain, it is genetically advantageous for breeding woodhoopoes to produce behaviorally variable young. The point can be illustrated by observations we began to make in 1977. That year three out of four male-sibling helpers in a flock composed of the usual breeding pair and five nonbreeders emigrated to a neighboring territory. There they drove out the

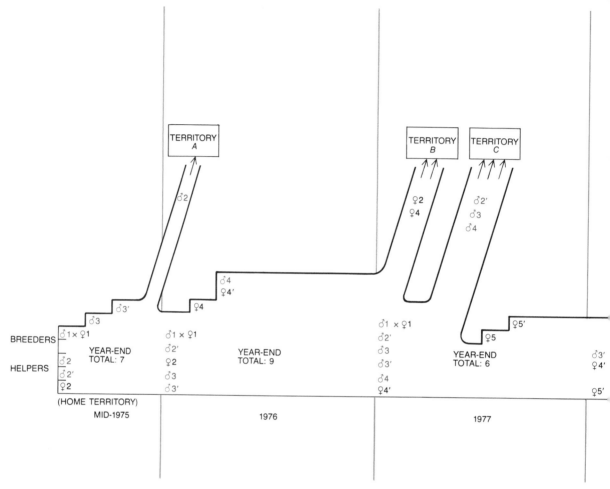

**Figure 9.5 SHIFTING MAKEUP of one woodhoopoe flock over a 66-month interval is shown in this diagram. When first banded, the flock had one breeding pair (*male 1 and* *female 1; the numbers identify them as the first observed* *generation*) plus three other birds (*males 2 and 2' and* *female 2*), probably offspring of the breeding pair (hence

lone male occupant and joined three females. The fourth male helper did not emigrate. Two years later the male parent died and the fourth male helper's mother and two sisters subsequently emigrated, leaving the male helper as the sole inheritor of his parents' territory.

This male soon mated with one of two female emigrants from a neighboring flock. In late 1979 and in 1980 the new breeding pair raised a total of seven offspring. In contrast to this record of reproductive success two of the male's three emigrant siblings had died by the end of 1979 without ever having produced young. The youngest of the three

was still alive in early 1982, but it has fledged only one offspring. If the fourth male sibling had not stayed at home, its contribution to the woodhoopoe gene pool might easily have been nil.

Although many emigrant groups consist of siblings, we also observed unrelated woodhoopoes of the same sex merging to provide the nucleus of a new flock. Typically this happens when an older, dominant bird allows a younger, subordinate individual to join it. We observed such behavior only when an older male needed an ally to defend and hold its territory. Two females would occasionally allow a third unrelated female to join them

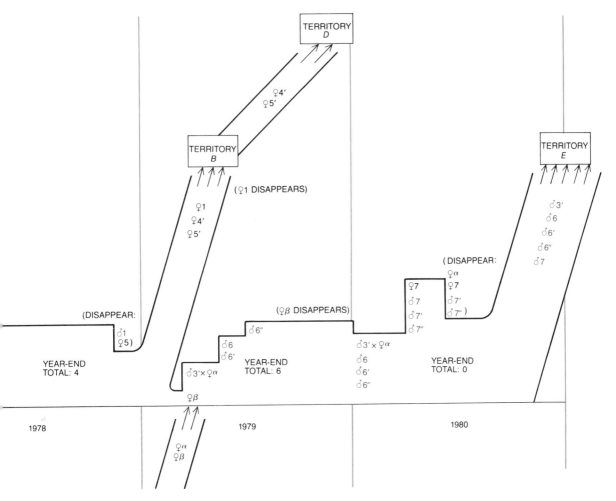

second generation). Over the years, group size rose after successful breeding seasons (e.g., three new offspring in 1976) and immigrations (*females α and β in 1979*), but fell after emigrations (every year except 1978) and disappearances (e.g., *breeding male 1* late in 1978). Emigrations typically involve two or more same-sex kin of different ages.

under the same circumstances. Of 40 observed instances of the formation of a new flock (or the replacement of all the members of one sex or all but one member of one sex) 17 of the groups included nonrelatives of the same sex and 17 included known or probable relatives of the same sex. In six instances kinship ties, if there were any, were not known to us.

Let us now suggest how cooperative breeding behavior may have originated among the green woodhoopoes and how this behavior in turn may have led to reciprocal assistance among unrelated individuals of the same sex. First, as we have seen, three features of the woodhoopoes' environment and life cycle appear to have set the stage for the development of cooperative breeding by favoring those breeding pairs that allowed at least some of their offspring to remain indefinitely in the parental territory. The three features are a high adult death rate, an unpredictable birth rate due to a fluctuating food supply and the uncertain availability of roost sites.

The extended term of residence for adult offspring confers two benefits on the breeding pair. The first of them is the opportunity for territorial

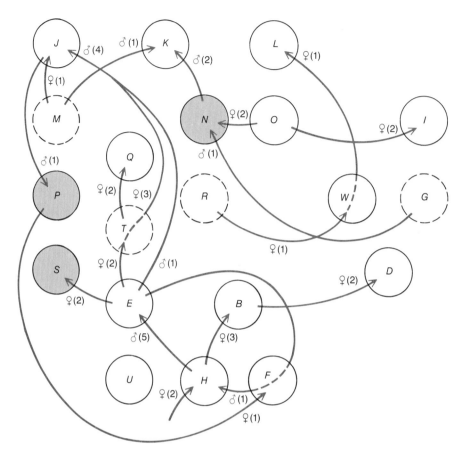

Figure 9.6 EMIGRANTS' MOVEMENTS among 20 wood-hoopoe territories during a four-year period (1978–81). Where the territories are the same as those in Figures 9.1 and 9.5 the identifying letters correspond. Thirty-one individual migrations are shown, involving flights by 14 males and 17 females. Birds that enter and leave one territory en route to another (for example from E to J by way of T) are counted only once. The same birds, however, may move more than once. For example, two of the three females that went from H to B subsequently moved from B to D. The four territories with broken perimeters were abandoned and three shown in color were colonized during this period.

expansion available to a flock that consists of more than two adults. In this uniform acacia-woodland environment a larger territory means an increased foraging area. Similarly, the presence of more than two adult birds means that territorial defense weighs less heavily on the individual adults.

The second benefit of an extended term of residence is whatever assistance the young adults may offer the breeding pair in feeding and protecting their nestlings. The presence of helpers probably reduces the cost to the breeding pair of annual nesting, thereby increasing the pair's potential lifetime reproduction record. Then when opportunities to

occupy new territory arise for the adult non-breeders, these emigrants find their chances of success enhanced by the availability of allies from among their younger flock mates of the same sex.

Once this interdependence for territorial defense and expansion, for reproduction and later for the acquisition of new territories became established among related individuals, any unaided woodhoopoe pairs or single birds were placed at an overwhelming competitive disadvantage. If most such birds are ever to gain territory for themselves and eventually to breed, the best option open to them once a mate is found is to obtain unrelated allies of

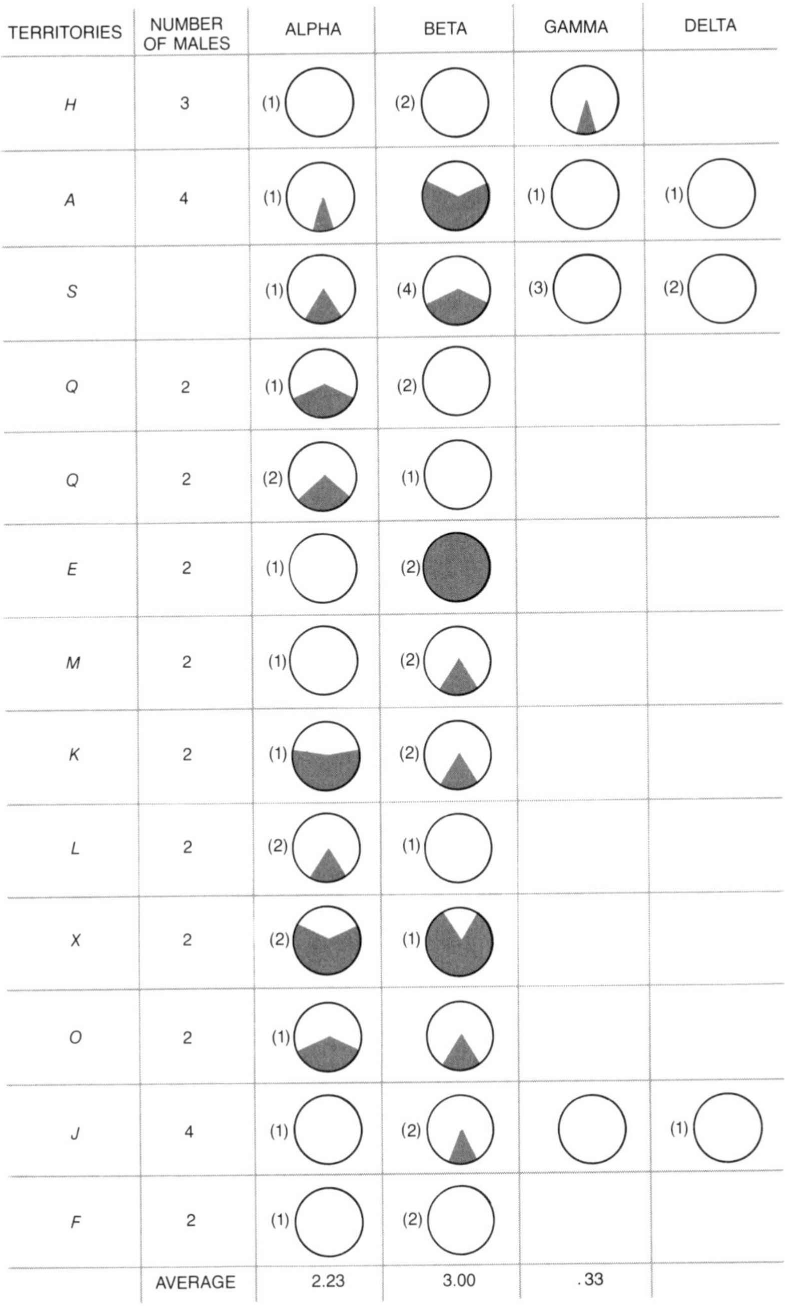

TERRITORIES	NUMBER OF MALES	ALPHA	BETA	GAMMA	DELTA
H	3	(1)	(2)		
A	4	(1)		(1)	(1)
S	2	(1)	(4)	(3)	(2)
Q	2	(1)	(2)		
Q	2	(2)	(1)		
E	2	(1)	(2)		
M	2	(1)	(2)		
K	2	(1)	(2)		
L	2	(2)	(1)		
X	2	(2)	(1)		
O	2	(1)	(2)		
J	4	(1)	(2)		(1)
F	2	(1)	(2)		
	AVERAGE	2.23	3.00	.33	

Figure 9.7 REPRODUCTIVE SUCCESS of emigrating male woodhoopoes is plotted. Fourteen groups of two to four males of varying ages (*alpha denoting the eldest, beta the second eldest, etc.*) left their home territories (*far left column*) and either failed to breed subsequently (*empty pies*) or achieved some success. The colored wedges are proportional to the greatest success observed, viz a total of eleven offspring sired by the beta male from territory *E*. Parenthetical numbers indicate the order of group-male deaths. Although alpha males were first to breed, beta males obtained greater average reproductive success.

Figure 9.8 NONBREEDING HELPERS from a flock of green woodhoopoes surround a fledgling (*center*) in this painting. Two have brought the main local foodstuff, caterpillars, to the juvenile woodhoopoe and one is grooming it. The three remaining helpers, their bills agape, are vocalizing in the direction of the fledgling. A woodhoopoe flock may include as many as 16 birds, but only one adult pair breed. The nonbreeding adults in the flock take part in the upbringing of the breeding pair's offspring and also assist in the defense of the flock's territory.

their own sex. This they do by allowing unaffiliated younger and subordinate birds to join them.

How well do the facts of the woodhoopoe social system correspond to the various theories that deal with cooperative behavior? Three theories are particularly relevant: Darwinian individual selection, kin selection and reciprocity. The theory of individual selection, as set forth by Charles Darwin, is based on the premise that animals behave in a way that maximizes opportunities for them to produce their own offspring. In the context of cooperative breeding among birds, a nest helper of either sex might eventually gain valuable resources such as territory, for example by first serving as an apprentice to the older, socially dominant breeders.

Kin selection, an extension of individual selection, is a theory developed (primarily by W. D. Hamilton of the University of Michigan at Ann Arbor) to account for what seemed to be altruistic behavior. In order to explain the apparent altruism seen in nature Hamilton proposed that if the unselfish behavior shown by an individual was directed toward a relative, it would serve to promote genes shared by the altruist and the recipient of the altruist's aid, so that the costs and benefits of such behavior should be correlated with the degree of genetic relatedness between the two interacting individuals.

The idea behind the third theory, reciprocity, is that an individual provides aid to another individual with the expectation that it will be repaid (at a value equal to or greater than that of the aid but not necessarily in the same form). Reciprocity is a specialized means of obtaining benefits for the individual and thus does not differ fundamentally from Darwinian individual selection.

Now, because the majority of the birds that interact within the woodhoopoes' closed social unit are relatives, it could be argued that what has been crucial to the evolution of the system is kin selection: natural selection where animals preferentially aid relatives, perhaps at a cost to themselves. Indeed, one kinship bond has clearly been fundamental to the evolution of the woodhoopoe social system: the bond between parent and offspring. It is not clear, however, whether other kinship ties are necessary to the maintenance of the system. For example, in three to four flocks each year since at least 1977 we have observed helpers that were not related to the nestlings they were tending. Some of these nonkin not only fed the nestlings but also fed the female member of the breeding pair while she was still incubating the eggs. Moreover, the merger of unrelated individuals either to increase the size of the flock or to form new flocks suggests that genetic relationships are not all-important to the woodhoopoes' forms of cooperation.

Reciprocity of various kinds occurs among flock mates, whether they are related or not, and it seems to be an important component of the woodhoopoe social system. As we have seen, nonbreeding members of the flock both serve as helpers and aid in territorial defense. In return these individuals can benefit in two ways. First, if the helper outlives the breeder of its own sex in the flock, it may inherit breeder status. Second, as a result of interaction with the developing young the helper wins potentially valuable allies in the flock in the event of emigration.

Moreover, when an older helper emigrates to take advantage of a territorial breeding vacancy, it is usually accompanied by one or more younger allies. By helping to establish the older bird's position in the new flock the younger birds can be thought of as repaying the older bird for the services they received in the nest. The younger birds will also tend the older bird's offspring, thereby establishing a kind of cross-generational reciprocity because the fledglings the older bird feeds may eventually help the helper to advance its own breeding program. The merger of unrelated adult birds is also suggestive of reciprocity; because of the high death rate this form of cooperation may often lead to an eventual breeding status for all involved.

Finally, what appears to be the primary underlying basis for the woodhoopoes' complex behavior is Darwinian individual selection: the maximizing of individual reproductive opportunities with respect to the other members of the population. The various woodhoopoe strategies I have described are most parsimoniously explained as behavior that has evolved to maximize these opportunities under an unusual and unpredictable set of circumstances.

AVIAN EVOLUTION

. . .

Introduction

In many American universities, ornithology is still offered as a separate course. At first this seems anachronistic, given that the biological principles students learn while studying birds are much the same as the principles that govern fish, centipedes, mammals and plants. Surely the retention of classical "-ology" courses is testimony only to curriculum inertia. While this may be part of the reason, there is also a persistent sense that students benefit from focusing, at least temporarily, on one reasonably discrete natural assemblage of life forms among the many in order to acquire an intuitive sense of the marvels of adaptive radiation (the evolutionary diversification of a single lineage into an amazing jumble of highly varied species and forms). Birds are often selected for this exercise because of their visibility and familiarity. (Besides, if the student decides to become a lawyer or short-order cook, bird-watching remains a viable avocation.)

Then too there is the troubling awareness that the biological diversity of our planet is imperiled, which may have dire and unforeseen consequences on both tangible and abstract levels. Species are lost, never to be regained, and no one will ever know what important secrets they may have taken with them. Too few citizens appreciate the exquisite uniqueness of each life form that would enable them to understand the stakes when human "progress" conflicts with preservation.

The concluding trio of chapters illustrates three extremes in the way scientists study phylogenetic questions: one deals in great detail with the single —quite possibly doomed—extant genus containing just three kiwi species; another is concerned with a single species that has been extinct for an unimaginably long time, and the last embraces all 9,000 living members of the class Aves simultaneously.

In Chapter 11, "The Kiwi," William A. Calder III examines a great many facets, ranging from the zoogeography of Gondwanaland to the oxygen physiology of single living eggs, before considering the ecological niches available in the absence of mammalian rivals in New Zealand. A dramatic glimpse of the problems these birds confront under the new rules recently came to light when it was discovered that a single German shepherd dog killed 500 of these flightless birds during a six-week rampage. Likewise, as shown in Chapter 12, "Archaeopteryx," by Peter Wellnhofer, the unique historical importance of the *Archaeopteryx* fossils, sitting virtually atop the avian family tree, helps make up for the paucity of material available for study a mere six specimens. That shortage, however, makes all the more impressive how much Wellnhofer and others have been able to deduce about the animal's life, even while debunking the charges of scientific fraud. On a much broader scale, Charles Sibley and Jon Ahlquist applied elegant new molecular biology techniques for decoding the phylogenetic information embedded in DNA to as many birds as possible (1,600 species from 168 families), as discussed in Chapter 10, "Reconstructing Bird Phylogeny by Comparing DNA's."

Whereas it is the similarities between *Archaeopteryx* and modern birds that help us learn from that comparison, it is the surpassing weirdness of kiwis that allows knowledge of their biology to contribute disproportionately to our appreciation of avian diversity. (In many ways, kiwis differ more from, say, crows than does the *Archaeopteryx* itself.) Finally, because the research strategies illustrated in these three chapters differ so markedly, we should note that the mix also provides a provocative glimpse of the adaptive radiation of Ornithologists.

Reconstructing Bird Phylogeny by Comparing DNA's

Differences between DNA's reveal evolutionary distances between species, making it possible to reconstruct and date the branchings of avian lineages and providing a basis for classifying living groups.

• • •

Charles G. Sibley and Jon E. Ahlquist
February, 1986

All organisms have ancestors; therefore all organisms have an evolutionary history. Because all the plants and animals presumably evolved from a single origin, they share a single phylogeny, or history. The reconstruction of this phylogeny is a primary goal of evolutionary biology. Living species are the topmost twigs of a vast phylogenetic tree whose larger branches and trunk are no longer directly visible. To reconstruct the tree of life it is necessary to determine the branching pattern and, if possible, date the branching events of the past.

Over the past 10 years we have used a technique that extracts evidence of phylogeny from the genetic material, DNA. The method, DNA-DNA hybridization, has enabled us to reconstruct the branching pattern of the major lineages of birds. The approximately 9,000 living species of birds are the descendants of lineages that began to diverge from one another about 150 million years ago in the late Jurassic and early Cretaceous periods, after the origin of birds from a reptilian ancestor. Pierce Brodkorb of the University of Florida has estimated that about 150,000 species of birds have existed.

The living species are only 6 percent of the total; the rest are extinct.

Our approach has been to measure the average difference between the DNA's of species representing the major groups of living birds and to use the results to reconstruct the branching sequence of the avian tree. The reconstructed phylogeny provides the basis for a classification of birds in which living species are assigned to taxonomic categories on the basis of their genealogical relationships. In some cases our results have indicated changes from traditional avian classifications.

The elements of a phylogeny are the branching pattern and the date of each branching event. A branching occurs when a barrier, usually a geographic one, divides a single species into two populations, which then diverge genetically and become the ancestors of two lineages. Each lineage in turn may split, and the process may repeat itself to produce a radiation of morphologically and ecologically varied species.

Until recently comparisons of the anatomical characters of living species were the only source of information about the pattern of branching. Such comparisons have answered many questions and have established the outlines of the history of life. Anatomical characters, however, are shaped by functional requirements; thus structure may provide false clues about phylogeny because the process of convergent evolution can produce similar structures in unrelated organisms. Swifts and swallows, for example, are superficially alike because both groups are specialized to feed on flying insects. In early classifications of birds the two groups were placed together. Later studies showed fundamental differences in anatomy, and it was eventually realized that the groups are not closely associated: swifts are distant relatives of hummingbirds, and swallows are related to other songbirds. Many cases of convergence are so subtle that they defy solution by anatomical comparisons.

To determine the dating of the branching events that were inferred from anatomical comparisons it has been necessary to rely on the fossil record. Partial phylogenies for certain groups have been reconstructed from fossil data, but the fossil record for some groups, including birds, is fragmentary. Moreover, although a dated fossil indicates approximately when the individual organism perished, the time its lineage diverged from that of its relatives usually remains uncertain.

Clearly a direct method for measuring genealogical distances among extant lineages and dating the divergences between lineages should improve the reconstruction of phylogenies. The genetic relationships among living species reflect their evolutionary history; because genetic change is mainly divergent, the genetic difference between any two lineages is related to the length of time since the lineages last shared a common ancestor. To reconstruct the phylogeny of birds we therefore studied their genetic material.

In all organisms except for certain viruses DNA is the genetic material. It is a double-stranded molecule, in which each strand is a sequence of four kinds of chemical units called nucleotides. Each nucleotide is composed of a five-carbon sugar, a phosphate group and a base. The nucleotides differ from one another only in their bases: adenine (A), thymine (T), cytosine (C) and guanine (G). The bases along the two strands form complementary pairs, held together by hydrogen bonds: an A pairs with a T and a C with a G. Genetic information is encoded in the sequence of bases. Specific sequences of bases form genes, which code for the many kinds of proteins that make up most of the structures of plants and animals and control their functions.

Within the genome, or complete set of genes, of a cell most genes are present as a single copy. Between about 3 and 5 percent of the different sequences in higher organisms occur as more than one copy in each genome. These repeated sequences may make up about 40 percent of the total volume of DNA in a cell.

There are about two billion pairs of nucleotides in the genome of a bird. The technique of DNA-DNA hybridization enables us to compare these huge numbers of genetic units and measure the genetic differences between living species. From these measurements we are able to reconstruct the branching pattern of the phylogeny and, by calibrating the measured genetic differences against time, calculate the approximate dates of the divergences between living lineages.

The technique of hybridization depends on the properties of DNA molecules. When a solution containing double-stranded DNA is heated to boiling, the hydrogen bonds between complementary bases "melt," or rupture, and the DNA separates into single strands. The hydrogen bonds are the weakest bonds in DNA, and the rest of the molecule is not damaged by boiling. As the melted sample cools, the single strands collide by chance. If colliding strands have complementary sequences of bases, they will reassociate into the double-stranded structure as complementary bases "recognize" each other and the hydrogen bonds between them are reestablished. If the reassociation takes place at a low temperature, the restored duplex, or double-stranded DNA, can contain numerous mismatches, but at a temperature of 60 degrees C. about 80 percent of the bases must be properly matched for the duplex to be stable. Under such conditions single strands of DNA will reassociate only with their complementary partners, and the original double-stranded DNA will be restored.

When the single-stranded DNA's from two different species are combined and incubated at 60 degrees C., hybrid double-stranded DNA will form only between homologous base sequences: sequences inherited from a common ancestor of the two species (see Figure 10.1). Only homologous sequences contain enough complementary pairs to form thermally stable duplexes at 60 degrees. A hybrid duplex of DNA from different species will contain mismatched bases because the two lineages have incorporated different sets of mutations since

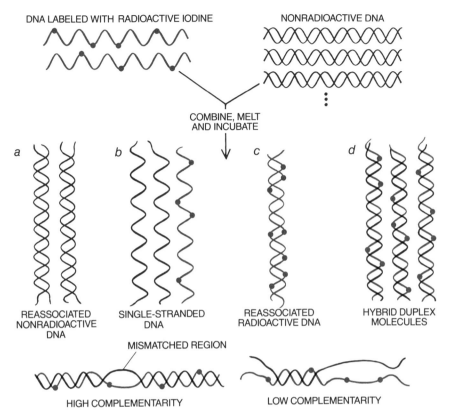

DNA LABELED WITH RADIOACTIVE IODINE   NONRADIOACTIVE DNA

COMBINE, MELT AND INCUBATE

a — REASSOCIATED NONRADIOACTIVE DNA

b — SINGLE-STRANDED DNA

c — REASSOCIATED RADIOACTIVE DNA

d — HYBRID DUPLEX MOLECULES

MISMATCHED REGION

HIGH COMPLEMENTARITY   LOW COMPLEMENTARITY

Figure 10.1 HYBRIDS ARE FORMED from a small amount of radioactive DNA plus 1,000 times as much unlabeled DNA from the same or different species (*top*). The DNA is melted into single strands then reassociated. Most unlabeled DNA binds harmlessly with other unlabeled strands (*a*). Some fail to reassociate (*b*) at all, and only about 1 percent of the radioactively labeled strands hybridize with other (rare) labeled strands (*c*). Some DNA forms hybrid duplex molecules, having one labeled and one unlabeled strand (*d*). In DNA hybrids from two species, the proportion of nucleotide bases matching complementary partners on the adjacent strand depends on the genetic similarity of the species. Well-matched hybrids (*bottom left*) are linked by more bonds, hence melt at higher temperatures than poorly matched ones (*bottom right*).

they last shared an ancestor. Thus an *A* may be opposite a *C* or a *G* may be opposite a *T*, and no bonds will form between the bases. Since the melting temperature of the duplex is proportional to the number of hydrogen bonds between the two strands, such mismatches will cause the hybrid DNA to melt at a temperature lower than that required to melt perfectly base-paired double strands.

In the DNA-DNA hybridization procedure DNA is extracted from the nuclei of cells and separated from the proteins and other cell constituents (see Figures 10.2 and 10.3). The long strands are sheared into fragments averaging 500 nucleotides in length. Most of the repeated gene sequences are removed from the DNA of the species that is to be compared with other species, and the "single copy" DNA that remains is labeled with radioactive iodine. A small quantity of the radioactive DNA (known as the tracer) is combined with a much larger amount of unlabeled DNA (the driver) from the same species. The same tracer is also combined with the driver DNA's of other species. Each mixture will yield a different hybrid: either a homoduplex, in which the tracer and driver species are the same, or a heteroduplex, in which the two strands represent different species. The homoduplex form provides a standard

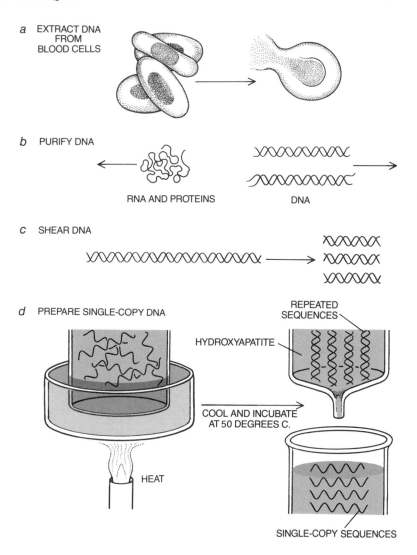

a EXTRACT DNA FROM BLOOD CELLS

b PURIFY DNA

RNA AND PROTEINS

DNA

c SHEAR DNA

d PREPARE SINGLE-COPY DNA

REPEATED SEQUENCES

HYDROXYAPATITE

COOL AND INCUBATE AT 50 DEGREES C.

HEAT

SINGLE-COPY SEQUENCES

Figure 10.2 AVIAN DNA-DNA HYBRIDIZATION studies start by extracting DNA from red blood cell nuclei (*a*). The DNA is cleaned (*b*) and cut into fragments (*c*). Boiling unzips the double-stranded DNA. As it cools, commoner sequences reassociate faster than single copies, which pass unbound when the mixture is passed through a hydroxyapatite column (*d*). The emerging single-copy DNA is labeled with radioactive iodine (*e*) and mixed with unla-

against which the melting properties of the heteroduplexes can be compared.

The tracer-driver mixtures are then boiled for five minutes to dissociate the double-stranded molecules into single strands. To allow the single strands to reassociate into duplexes they are incubated for 120 hours at 60 degrees C. in a sodium phosphate buffer solution. Each of the different double-stranded hybrids that results is placed on a column of hydroxyapatite, a form of calcium phosphate that binds double-stranded DNA but not the single-stranded form. The columns are immersed in a water bath at 55 degrees C., and the temperature of the bath is raised in increments of 2.5 degrees to 95 degrees. At each of the 17 temperatures the single-stranded DNA fragments produced by the melting of the duplexes are washed from each column into a vial. The radioactivity in each vial is counted, indicating how much of the corresponding hybrid had melted. The results are plotted as a melting curve: a graph

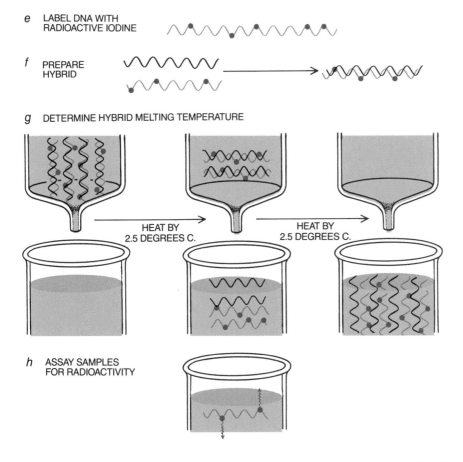

e   LABEL DNA WITH RADIOACTIVE IODINE

f   PREPARE HYBRID

g   DETERMINE HYBRID MELTING TEMPERATURE

HEAT BY 2.5 DEGREES C.

HEAT BY 2.5 DEGREES C.

h   ASSAY SAMPLES FOR RADIOACTIVITY

beled DNA from the same species or a different one. After incubation DNA hybrids form, containing one labeled and one unlabeled strand (*f*). Back in the column the hybrid is heated in increments; at each increment the single strands released by the hybrid melting wash from the column into a vial (*g*). Radioactivity indicates how much hybrid melted at each temperature (*h*).

showing how much of the hybrid had melted at each temperature (see Figure 10.4). The median difference in degrees Celsius between the homoduplex curve and each heteroduplex curve is a measure of the median genetic difference between the tracer species and each driver species it was compared with.

The difference between the DNA's of two species can serve as an indicator of the genealogical distance between them only if DNA can be assumed to change at an average rate that is the same in all lineages. Emile Zuckerkandl and Linus Pauling, then at the California Institute of Technology, proposed in 1962 that proteins evolve at constant rates, and "molecular clocks" have been discussed and debated ever since.

We have found that the DNA clock seems to tick at the same average rate in all lineages of birds. The evidence comes from a procedure known as the

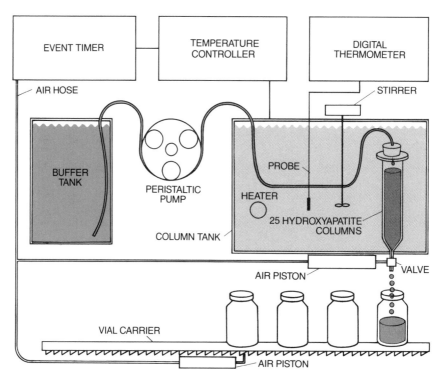

Figure 10.3 APPARATUS FOR DNA-HYBRID ANALY-
SIS, the DNALyzer, processes 25 hybrids simultaneously.
An event timer coordinates the whole operation and a
controller regulates the temperature of the water bath in
which the hybrid DNA samples, contained in a row of 25
hydroxyapatite columns, are immersed. The timer's compressed air drives a carrier to move the vials into position
and opens the valves at the bottom of the columns, allowing pumped buffer solution to flow through the columns.
The buffer flushes the single-stranded DNA from the
melting hybrids, into the vials. The valves then close and
the heater raises the water temperature by 2.5 degrees C.
before the next cycle begins.

relative-rate test (see Figure 10.5) which was suggested in 1967 by Vincent M. Sarich and Allan C.
Wilson of the University of California at Berkeley. A
relative-rate test compares any three species of
which two are known to be more closely related to
each other than either one is to the third. If we
choose such a trio of bird species and compare
tracer DNA from the outlying species with driver
DNA from each of the other two, we find that the
genetic distances between the outlier and each of
the other species (indicated by the melting temperatures of the DNA hybrids) are always equal, within
the limits of experimental error. Since the same
length of evolutionary history separates both of the
driver species from the last ancestor they shared
with the tracer species, the DNA of both driver
species must have changed at the same average rate.
Our reconstructed phylogeny of living birds includes thousands of such trios of species that yield
the same result and attest to the uniform average
rate of the DNA clock in birds.

This apparent constancy may seem to be magic
(or nonsense) at first glance, but it may be simply
the result of measuring differences between sequences of billions of base pairs after millions of
years of evolution. Natural selection dictates that
different genes evolve at many different rates, and
any individual gene may evolve at different rates at
different times, but the range in the rates of all
genes is narrow and the number of genes in the
genome of a bird is huge. As the rate of evolution of
one gene speeds up, another gene may be statistically likely to slow down by the same amount. The
balancing of the rates of change in different genes
need not occur simultaneously; the apparent constancy arises over millions of years. Whatever the

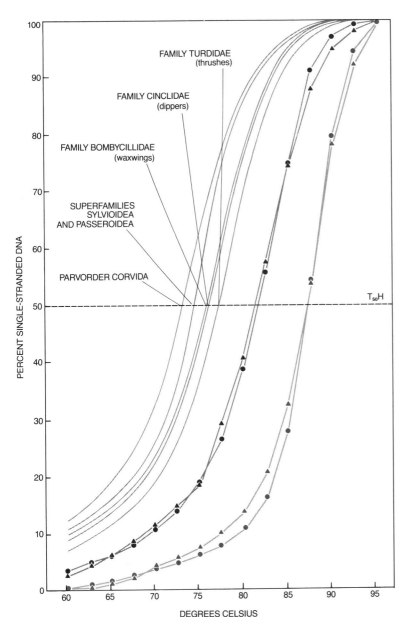

PERCENT SINGLE-STRANDED DNA

FAMILY TURDIDAE
(thrushes)

FAMILY CINCLIDAE
(dippers)

FAMILY BOMBYCILLIDAE
(waxwings)

SUPERFAMILIES
SYLVIOIDEA
AND PASSEROIDEA

PARVORDER CORVIDA

$T_{50}H$

DEGREES CELSIUS

Figure 10.4 MELTING CURVES OF DNA HYBRIDS indicate DNA differences between starlings and mockingbirds and between these birds and other lineages. The horizontal axis shows temperatures to which hybrid samples were heated; the vertical, how much DNA melted at each temperature. Colored curves show baseline values, when both of the "hybrid" samples came from the same species, either a mockingbird (*dots*) or starling (*triangles*). The unlabeled black curves represent hybrids between these two species (different families). The remaining smooth curves are for hybrids between one of these standards and various (indicated) comparison groups. The temperature at which half of the hybrid DNA melted is a useful scale: the lower the T_{50} H the more distantly the species are related.

correct explanation may be, avian DNA appears to evolve at a uniform average rate.

Because mismatched bases in DNA hybrids are the result of genetic changes that have been fixed in the two lineages since they last shared a common ancestor, the number of mismatches is proportional to the time the lineages have been diverging. The median melting temperature of a DNA-DNA hybrid is therefore an indirect measure of the length of time since the branching occurred. By correlating the median melting temperature with a dated geologic event that caused an ancestral species to be divided into separate lineages, we can calibrate the DNA clock in absolute time.

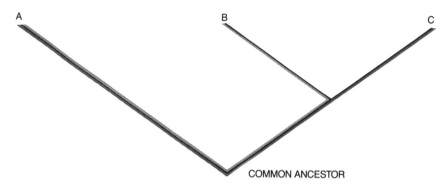

Figure 10.5 RELATIVE-RATE TEST confirms that DNA evolves at the same average rate in different species. The test can be done with any trio of species in which two (*B, C*) are more closely related to each other than either is to a third (*A*). If DNA from each of the two allied species is hybridized with DNA from the outlying species, the melting temperatures of the two hybrids will be identical, indicating that the hybrid DNA's contain equal numbers of mismatched bases. Thus the genetic distances between species *A* and *B* (*color*) and between species *A* and *C* (*gray*) are the same. Since the same length of time separates species *B* and *C* from the last common ancestor they shared with species *A*, the average rates at which their DNA has changed must be identical.

We have assumed, for example, that the common ancestor of the ostrich of Africa and the rheas of South America ranged across the protocontinent of Gondwanaland before continental drift split the protocontinent into the southern continents of today and opened the Atlantic Ocean during the Cretaceous period. The geologic evidence indicates that the Atlantic became a barrier for flightless animals about 80 million years ago. Thus the Ostrich and rhea lineages must have diverged about then. Dividing 80 million years by the difference between the median melting temperature of ostrich/rhea DNA heteroduplexes and that of ostrich/ostrich or rhea/rhea homoduplexes yields a calibration constant, in millions of years of divergence per degree of reduction in median melting temperature.

We have seven similar dated divergences between bird lineages caused by three geologic events, two of them about 80 million years ago and one some 40 million. Each instance yields a calibration constant of between 4.3 and 4.7, with the average at 4.5. Hence a median reduction in melting temperature of one degree C. is equivalent to about 4.5 million years since the two lineages shared their most recent common ancestor. This constant is tentative and subject to correction, but we use it to calculate approximate divergence dates.

During the past 10 years we have made more than 25,000 DNA-DNA hybrid comparisons, using genetic material from about 1,600 species, which together represent 168 of the 171 traditional families of living birds. From these data we have reconstructed the phylogeny of most of the groups of birds. The examples that follow illustrate some of the problems we have examined and the solutions indicated by the DNA hybridization data.

The barbets are small and usually brightly colored birds with tufts of bristles at the base of their relatively large bills. The Old World barbets live in Africa and southern Asia; the New World species live in the tropics of Central and South America. Traditionally the barbets have been placed in the family Capitonidae and were thought to be related to the woodpeckers (the Picidae) and to the large-billed, fruit-eating toucans (the Ramphastidae) of the New World tropics. Many taxonomists have remarked that the small species of toucans and the large species of New World barbets show similarities in morphology, and two recent studies have emphasized the close relationship between the barbets and the toucans (see Figure 10.6). In 1984 Philip Burton of the British Museum of Natural History concluded from a study of the head region: "It seems reasonable simply to regard toucans as a specialized group of barbets which have arisen and radiated in South America." Similarly, in 1985 Lester L. Short, Jr., of the American Museum of Natural History stated: "Toucans effectively are large, specialized, toothbilled barbets."

The DNA comparisons agree with Burton and Short and enable us to add the dimension of time

EMERALD TOUCAN (NEW WORLD)

TOUCAN BARBET (NEW WORLD)

BLACK-COLLARED BARBET (OLD WORLD)

Figure 10.6 BARBET AND TOUCAN SPECIES perch on a branch symbolizing the genealogical relationships of the groups to which they belong. The barbets of the New World and those of Africa and Asia have traditionally been placed in one family and the toucans in another. Comparisons of DNA's showed that in spite of the birds' appearance the New World barbets are related more closely to the toucans than to the Old World barbets. The degrees of difference between DNA's indicated the African barbets last shared an ancestor with the New World barbets and the toucans about 55 million years ago; the latter groups diverged about 30 million years ago.

to their conclusions. The DNA indicates that the branching between the Old World and the New World barbets took place about 55 million years ago. The toucans branched from the New World barbet lineage more recently, about 30 million years ago. Hence the toucans are more closely related to the New World barbets than the two groups of barbets are to each other. To reflect these relationships our classification of these groups and the others belonging to the same order looks like this:

Order Piciformes
    Parvorder Picida
        Family Picidae (woodpeckers)
        Family Indicatoridae (honey-
            guides)
    Parvorder Ramphastida
      Superfamily Megalaimoidea
        Family Megalaimidae (Old
          World barbets)
      Superfamily Ramphastoidea
        Family Ramphastidae
          Subfamily Ramphastinae
          (toucans)
          Subfamily Capitoninae (New
          World barbets)

Our classification is based on the branching pattern of the phylogeny and the times of origin of the groups, as dated by the DNA comparisons. We divided the evolutionary time scale into segments of 10 million years and assigned a taxonomic category to each segment. Orders, in our scheme, are those lineages that branched from other lineages 90 to 100 million years ago and suborders are those that branched 80 to 90 million years ago. Infraorders originated 70 to 80 million years ago, parvorders 60 to 70, superfamilies 50 to 60, families 40 to 50, subfamilies 30 to 40 and tribes 20 to 30 million years ago. One result of the procedure is to make groups at the same categorical level approximately equal in their degree of evolutionary divergence. The boundaries we have constructed are not rigid, but they provide the basis for a classification that reflects the DNA-derived phylogeny and approaches the ideal of equivalent categories.

There is controversy, however, over how to classify organisms; some taxonomists prefer to base categorical rank on their evaluation of the degrees of morphological specialization of groups. On that basis the two groups of barbets might be placed in the same family and the toucans in an adjacent

family or superfamily. Such a classification recognizes the birds' distinctive appearances but conceals their phylogenetic relationships. In one traditional arrangement, currently in wide use, the toucans, barbets and some other groups are classified as follows:

Order Piciformes
  Suborder Galbulae
    Superfamily Galbuloidea
      Family Galbulidae (jacamars)
      Family Bucconidae (puffbirds)
    Superfamily Capitonoidea
      Family Capitonidae (barbets)
      Family Indicatoridae (honey-
        guides)
    Superfamily Ramphastoidea
      Family Ramphastidae (tou-
        cans)
  Suborder Pici
    Family Picidae (woodpeckers)

Not only is this classification at odds with the DNA data on the toucans and barbets, but also the DNA indicates that the jacamars and puffbirds of the New World should be placed in a separate order, not in the Piciformes. There is no sign that the debate about how to classify plants and animals will be resolved soon.

The vultures of the Old World are closely related to the hawks and eagles. The New World vultures, including the condors, the Turkey Vulture and the Black Vulture, superficially resemble the Old World vultures. Both groups are carrion eaters, and they have usually been placed together in the order Falconiformes: the diurnal birds of prey. The New World vultures share many morphological traits with the storks, however, and some taxonomists have argued that the New World vultures and the storks belong in the same order (see Figure 10.7). This arrangement was proposed by Alfred B. Garrod in the 1870's and by David Ligon, then at the University of Michigan, in 1967, but recent classifications have ignored the evidence of a condor-stork alliance and continued to place the condors and their relatives in the Falconiformes.

The DNA comparisons support the morphological indications suggesting the New World vultures and the storks are each other's closest living relatives; the data show the groups diverged from a common ancestor about 35 to 40 million years ago. Similarities between the vultures of the New and

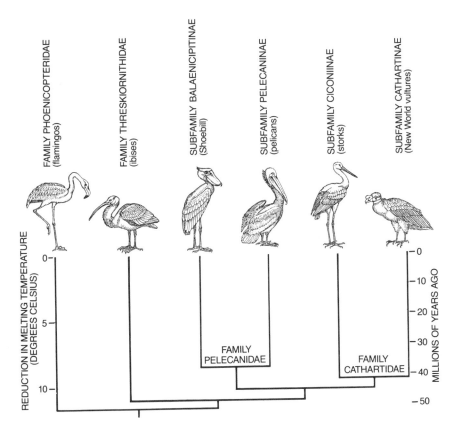

Figure 10.7 NEW WORLD VULTURES AND STORKS are each other's closest living relatives, according to the phylogeny reconstructed from DNA comparisons. Superficially the New World vultures (which include the condors) resemble the vultures of the Old World. The similarities must be due to convergent evolution; the Old World vultures belong to a different cluster of lineages. DNA studies also showed that the pelicans and the Shoebill, an African species, are the nearest neighbors of the storks and condors in the phylogenetic tree. In addition the DNA confirmed the close alliance between the Shoebill and the pelicans.

Old worlds are the result of convergent evolution related to their carrion-eating habits.

The totipalmate birds are those with webs between all four toes: pelicans, cormorants, anhingas, boobies, gannets, frigatebirds and tropicbirds. All except the tropicbirds also have an obvious gular pouch, or throat pouch, between the branches of the lower mandible; in tropicbirds the gular pouch is small and concealed. Mainly because of these two shared characters the totipalmate birds have usually been grouped together as members of the order Pelecaniformes, although it has sometimes been suggested that tropicbirds and frigatebirds are related to some other group.

Another bird, the Shoebill, has been proposed as a member of the same cluster. The Shoebill, a large storklike bird with a huge bill, lives in the swamps of eastern Africa and feeds on lungfishes and other aquatic prey. The Shoebill has usually been viewed as a relative of the storks or the herons, but a 1957 study of its skeleton led Patricia A. Cottam, then at the British Museum, to conclude that the Shoebill is most closely related to the pelicans. Cottam's evidence was dismissed by most taxonomists as the result of convergent evolution, but our DNA comparisons support her. The divergence between the Shoebill and the pelicans took place about 35 to 40 million years ago. The nearest cluster to the pelican-Shoebill group turns out to be the group that includes the storks and the New World vultures. The two groups diverged 40 to 45 million years ago.

The surprises were not over. The DNA also showed that the traditional order Pelecaniformes, in which the totipalmate birds had been placed, is a polyphyletic group: it is composed of several subgroups that are more closely related to other birds than they are to one another. The pelicans are most closely related to the Shoebill, storks and condors. The cormorants, anhingas and boobies are allied with one another and seem to be distantly related to the herons; the frigatebirds are relatives of the tubenosed seabirds (the albatrosses and petrels), and the tropicbirds appear to represent a separate lineage with no close living relatives. Hence the totipalmate foot and the gular pouch either evolved by convergence in separate lineages or were inherited from a distant common ancestor of several groups of birds. In descendants that do not show these characters their genetic basis may have been repressed. Such dormant genes are known in some groups of birds and in other animals.

Polyphyletic clusters of bird lineages have been revealed in the past. In some early classifications the birds with palmate feet (feet with webs between the three front toes only) were grouped together. It was soon realized that not all palmate birds, which include the ducks, albatrosses, penguins, loons, gulls and auks, are closely related, and they were assigned to several different groups. In the case of the totipalmate birds, however, with the gular pouch as supporting evidence, it seemed impossible that they did not belong to a monophyletic cluster—one in which all the members share the same most recent common ancestor.

In a recent study Joel Cracraft of the University of Illinois Medical Center at Chicago compared 45 skeletal and seven behavioral characters among the totipalmate birds and the penguins, loons, grebes, albatrosses, petrels and the Shoebill. He found 12 traits supporting his hypothesis that the totipalmate species form a monophyletic group and six suggesting that the albatrosses and petrels are the sister group, or companion lineage, of the totipalmate birds. Among the totipalmate species, the tropicbirds emerged as a lineage distinct from the other totipalmate birds. Of the others, the frigatebirds appeared to be the descendants of the oldest branch, followed in order of branching by the pelicans, boobies, anhingas and cormorants.

Cracraft rejected the Shoebill as a relative of the totipalmate birds and ascribed its similarities to the pelicans to convergence. There are obvious disagreements between the DNA comparisons and the morphological evidence Cracraft used, but there are also several congruencies: the relationship between the totipalmate species (the frigatebirds in particular) and the albatrosses and petrels, the outlying position of the tropicbirds and the close relationships among the boobies, anhingas and cormorants. The major disagreements concern the relationship between the pelicans and the Shoebill and the position of the pelicans with respect to the other totipalmate birds.

It is unlikely that the DNA-based evidence of the polyphyly of the order Pelecaniformes will be accepted by most ornithologists soon. Nevertheless, we predict that appropriate comparisons will reveal morphological evidence that, like Cottam's 1957 study, is consistent with the DNA data.

The sandgrouse are pigeon- or ploverlike birds of the arid regions of Africa, Asia and southern Europe. Their relationships to other birds have been debated for more than a century. Are they related to the pigeons, to the plovers or to the galliform birds (chickens, pheasants and their relatives)? Each group has had proponents; most of the recent participants in the dispute support the pigeons or the plovers.

The DNA results are quite clear: sandgrouse are the sister group of a large part of the order Charadriiformes, including the sheathbills, thick-knees, plovers, oystercatchers, avocets, stilts, gulls, auks and coursers. The sandgrouse and their sister assemblage are in turn most closely related to the sandpipers and their allies. Hence the sandgrouse are neither pigeons nor plovers but are closer to the plovers than they are to the pigeons; similarities between the sandgrouse and the pigeons must be due to convergent evolution.

About 5,300 of the 9,000 species of living birds belong to the order Passeriformes; it includes the flycatchers, warblers, thrushes, sparrows, starlings, wrens, swallows, larks, crows and other species, generally small. Most of the passerine birds of South America are members of the suborder Oligomyodi, also known as the suboscines. The structure of the syrinx (the vocal apparatus) and other anatomical characters distinguish the suboscines from the oscines, or songbirds, which make up the suborder Passeres (the other branch of the Passeriformes). The New World suboscines evolved in South America while it was isolated from the rest of the world as an island continent from about 80 million years ago in the late Cretaceous period until about five million years ago.

One evolutionary lesson taught by the New World suboscines concerns the tropical antbirds, a group of about 235 species that have traditionally been placed in the family Formicariidae. In the early 1960's Mary Heimerdinger Clench and Peter L. Ames, both at Yale University, found that some antbirds have two deep notches in the posterior edge of their sternum, or breastbone; the other species have four notches. Ames also found that the two groups differ in the musculature of the syrinx. The 185 two-notched species occupy a range of habitats, but the 50 four-notched species are long-legged, short-tailed ground dwellers. Clench and Ames also pointed out that four-notched sterna occur in two other groups of birds: the tapaculos (family Rhinocryptidae) and the gnateaters (family Conopophagidae). Clench and Ames suggested that the ground-dwelling antbirds are more closely related to those groups than they are to the other antbirds.

The DNA revealed the same alliances. The comparisons showed that the two-notched antbirds branched from the four-notched lineage before the four-notched antbirds diverged from the tapaculos and the gnateaters. Thus the morphological and molecular evidence agreed, and the DNA data provided the branching order and the approximate dates of the divergences.

The other branch of the passerines, the suborder Passeres, includes about 4,000 of the 5,300 passerine species. The DNA comparisons revealed that the Passeres consists of two major groups, which we call the parvorders Passerida and Corvida (see Figure 10.8). These two lineages diverged from a common ancestor about 55 to 60 million years ago. The evidence shows the Passerida evolved in Africa, Eurasia and North America and the Corvida in Australia.

From about 60 to 30 million years ago, during the early and middle Tertiary period, Australia was isolated form other landmasses. The Corvida evolved many morphologically and ecologically specialized forms, including warblers, flycatchers, creepers, thrushes, babblers and nectar feeders—forms much like those to which the Passerida gave rise in other parts of the world. The birds of Australia were discovered and named, however, after European ornithologists had already classified those of most of the rest of the world. The Australian passerines

seemed to fit into categories that had been founded on specimens from other areas. Thus the Australian warblerlike passerines were assigned to the Sylviidae (which includes the true warblers), the Australian flycatchers to the Muscicapidae (the Afro-Eurasian flycatchers), and the treecreepers to the Certhiidae (the Eurasian-American creepers). The sitellas, nuthatchlike birds of Australia, were placed in the Sittidae (the family of true nuthatches) and the Australian honeyeaters were grouped with the superficially similar nectar-eating Afro-Asian sunbirds.

When we compared DNA's among various Australian passerines and between Australian species and their supposed relatives from Africa, Eurasia and North America, we found that the Australian endemics are more closely related to one another than they are to their morphologically similar counterparts from other continents. Convergent evolution had produced similarities between unrelated species of the two parvorders, and museum taxonomists had assembled the species into polyphyletic clusters containing members of both the Corvida and the Passerida. The same mistake had been made for most of the 400 species of passerines in Australia and New Guinea today. Many of the convergences are so subtle that the true relationships of the Corvida and the Passerida probably could not have been resolved from anatomical comparisons alone.

One result of the confusion was to conceal the fascinating story of the phylogeny of the Corvida, which parallels that of the marsupials. Both groups evolved while Australia was isolated; like the Australian passerines, the marsupials radiated into many of the same ecological niches occupied in Africa, Eurasia and North America by another group, in this case the placental mammals. In the process some of the marsupials assumed forms similar to those of other mammals. Unlike the Australian passerines, however, the marsupials were not confused with their counterparts elsewhere because they have a marsupium, or pouch, and other distinctive traits.

The Corvida of Australia produced the ancestors of a few groups that were able to emigrate to Asia as Australia drifted northward during the Tertiary period. Among the groups whose ancestors evolved in Australia is the tribe Corvini, which includes the same species as the traditional family Corvidae: the crows, ravens, jays, magpies and their relatives.

Today there are 23 genera of the Corvini; of these

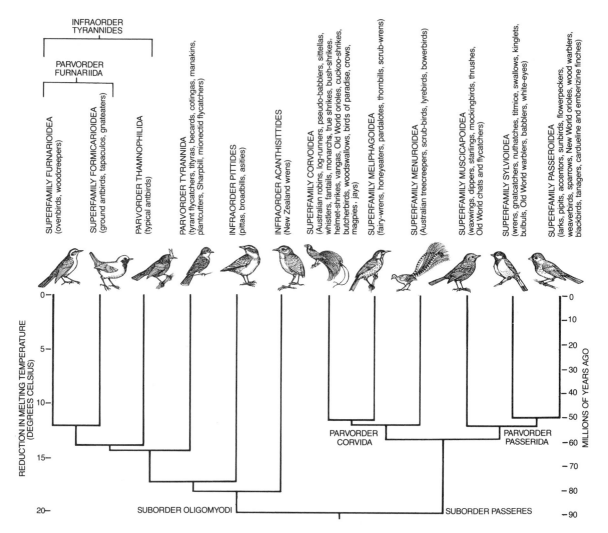

Figure 10.8 PHYLOGENY OF PASSERINE BIRDS, which include 5,300 of the 9,000 species of living birds, was reconstructed from DNA comparisons. For each bifurcation of the tree, the scale at the left shows the number of degrees by which the median melting point of hybrids formed of DNA's from species representing the two lineages is lowered with respect to perfectly matched DNA hybrids; the scale at the right dates the branchings. The reconstruction affects traditional classifications of living birds. In the suborder Passeres the results delineated two distinct groups: the parvorders Corvida and Passerida. The Corvida originated in Australia, although some groups in the Corvida are now distributed throughout the world.

15 occur in Eurasia and 10 in North America. Only three genera are found in Africa, and in South America the tribe is represented only by two genera of jays. The numbers reveal the chronology and pattern of the dispersal from Australia (see Figure 10.9). The earliest radiation occurred in southeastern Asia, and members of the lineages later extended their ranges to Europe, Africa and North America. South America was the last continent to be invaded. It was isolated from North America until between three and five million years ago; the two genera of jays that now occur in South America apparently expanded their ranges from the north after the emergence of Central America provided a land connection between the continents. The crows and ravens of the genus *Corvus* probably originated

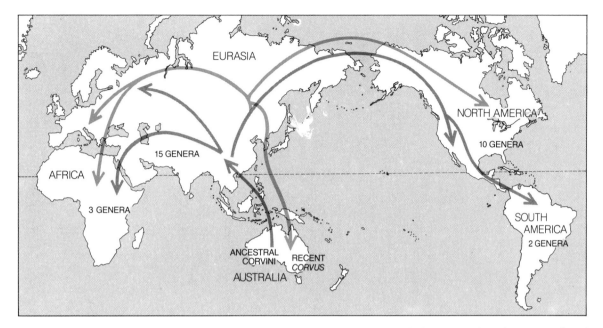

Figure 10.9 DISPERSAL OF TRIBE CORVINI, which includes crows, jays, magpies and their relatives, is mapped (*gray arrows*). DNA-DNA hybridization identified the Corvini as part of a larger group that evolved in Australia. The number of genera belonging to the Corvini on each continent reflects the order in which the continents were invaded after ancestors of the family reached Asia about 30 million years ago. The largest number of genera are found in Asia and the fewest in South America, which the Corvini reached only from three to five million years ago. The crows and ravens (genus *Corvus*) evolved in Asia and later spread to Africa and North America (*colored arrows*), colonizing Australia only in the past 100,000 years or so.

in Eurasia and spread nearly worldwide (except to South America), eventually even colonizing Australia.

Among the many surprising results from the comparisons of DNA's none was less expected than the discovery of the close relationship between two other groups of passerines: the starlings, which are native to the Old World, and the mockingbirds and thrashers of the New World (see Figure 10.10). The starlings have usually been considered related to the crows, and the mockingbirds were correctly placed near the thrushes. If the starlings were close relatives of the crows, they would be members of the Corvida, but the DNA clearly identifies the mockingbirds as members of the Passerida. Thus, if the traditional classification were correct, the starling and mockingbird lineages would have diverged between 55 and 60 million years ago.

The DNA comparisons revealed, however, that the two lineages diverged about 25 million years ago. Other work supports this close starling-mockingbird alliance: immunological comparisons of muscle proteins done in 1961 by William B. Stallcup, Jr., of Southern Methodist University, studies of the anatomy of the head region done in 1953 by William J. Beecher, then at the University of Chicago, and comparisons of the syrinx made by Wesley E. Lanyon of the American Museum of Natural History. It may also be significant that some starlings—myna birds, for example—are, like mockingbirds, excellent mimics. Even the Common Starling mimics the songs of other birds.

The close relationship between the starlings and the mockingbirds may reflect the history of climatic change in the Northern Hemisphere. During the early and middle Tertiary period, 65 to 30 million years ago, the climate of the Arctic was temperate; broad-leaved trees grew in northern Canada and Greenland. It seems probable that the common ancestor of the starlings and the mockingbirds was widely distributed over those regions, which served as a bridge between the Old World and the New World. Evidence from plant fossils indicates that the climate grew colder beginning about 30 million years ago, and the ancestral populations presum-

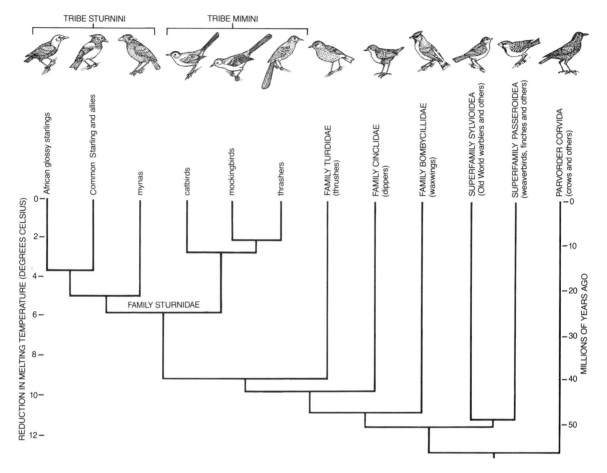

Figure 10.10 STARLINGS AND MOCKINGBIRDS occupy adjacent branches of the avian phylogenetic tree, according to data from DNA-DNA hybridization. The starlings (the tribe Sturnini) had been seen as related to the crows; the mockingbirds (the tribe Mimini) had been placed near the thrushes. If that classification were correct, the ances- tors of the mockingbirds and the starlings would have diverged almost 60 million years ago. Hybrid melting- point data showed instead that the starlings and mocking- birds are each other's closest relatives, having diverged about 25 million years ago. Both groups, which make up the family Sturnidae, are related to the thrushes.

ably moved southward. About 25 million years ago the populations in America and Eurasia became separated and began to diverge.

These are just a few of the discoveries DNA-DNA hybridization has made possible. The DNA comparisons present us with new hypotheses of avian relationships. If the DNA-based phylogeny of birds is closer to the one true phylogeny, it will be congruent with evidence from other sources. The results we have so far obtained agree with geologic history better than many earlier proposals about avian phylogeny, and the relationships indicated by

the DNA data are usually supported by at least some anatomical characters. We believe some aspects of morphology will prove to be congruent with the DNA evidence in all cases.

Yet DNA and the morphological characters tradi- tionally used to reconstruct phylogeny serve to pro- vide different kinds of information. Morphology shows how natural selection has modified structure to adapt organisms to the environment, whereas DNA comparisons give a direct indication of the branching pattern and the approximate branching dates among living lineages. Morphology is func- tional; the DNA clock keeps time.

The Kiwi

New Zealand, where this flightless bird lives today, had no mammals for 80 million years. In filling ecological niches that would have been occupied by the mammals the bird evolved mammalian characteristics.

· · ·

William A. Calder III
July, 1978

The animal kingdom is so rich in its diversity that evolution can appear to have been a random process. Consider the famous flightless bird of New Zealand, the kiwi. Natural selection is partly a matter of chance, affected by random mutations and historical accidents of parentage and location. At the same time it operates under certain constraints. For example, physical laws determine how large a bird can be and still fly. In their short life span human beings are not likely to witness the interaction between environment and heredity that gives rise to an organism adapted to an available environmental niche. The natural history of biological isolation, however, presents some suggestive case histories. The kiwi's is one of the most fascinating.

One often hears that birds exhibit marvelous evolutionary adaptations. The examples given may be the arctic tern's long migration, with its remarkable feats of navigation; the wingspan of those master soarers, the great albatrosses; the coevolution of the tiny hummingbirds and plants that bear red tubular flowers, and the elaborate mating displays of the bowerbirds. Can the same possibly be said for a bird that has lost its ability to fly, that probes for worms, that burrows in the dirt to lay a ridiculously over-

sized egg and that is known to nonbiologists mainly as the label of a brand of shoe polish? My answer, given the context of isolation and ancestry, is a clear yes.

There are three species of kiwi. The common or brown kiwi (*Apteryx australis*) is found on New Zealand's North Island and South Island and also on Stewart Island, off the southernmost tip of South Island (see Figure 11.1). The little spotted kiwi (*A. oweni*) was until recently present on both North Island and South Island but now is found only on South Island. The great spotted kiwi (*A. haasti*) is also found only on South Island. These birds are known nowhere else in the world.

Kiwis are the smallest members of the seven families of ratite birds: nonflying birds that lack the keeled breastbone to which are attached the massive pectoral muscles that flap the wings of flying birds. ("Ratite" is from the Latin *ratis*, raft, an unkeeled vessel.) Among the other ratites are the ostrich, the emu and the extinct giant moa of New Zealand (see Figure 11.2). In addition to lacking a keeled breastbone the kiwi shares with the other ratites an adult plumage that is the same in structure as the juvenile down of other birds. The kiwi's abundant feathers are long and flexible; they lack

Figure 11.1 THE COMMON OR BROWN KIWI (*Apteryx australis*) is the principal species of three still present in New Zealand. It has a shaggy appearance because its plumage resembles the juvenile down of other birds. It feeds largely by probing for worms with its long bill.

the usual interlocking mechanism, giving the plumage a shaggy, furlike appearance. The feathers do not ruffle as the kiwi digs the burrows where it spends the daylight hours, and they provide insulation both in the cool, damp underground habitat and in environments that range from semitropical in the far north of North Island to virtually subpolar in the Southern Alps of South Island.

The kiwi lacks an external tail, and its wings are reduced to vestiges that are hidden under its plumage. Its sense of touch is augmented by well-developed rictal bristles at the base of the bill and a sensitive region at the tip. Its legs are very strong; they serve the bird not only for locomotion but also for burrowing and fighting. The kiwi's posture when it is digging resembles the attitude of a swimming duck.

The kiwi feeds on insects, berries, seeds and other plant materials, but its principal food is the earthworm. New Zealand has a particularly rich earthworm fauna, and the kiwi has 178 native and 14 introduced species of worms at its disposal. The bird's foraging technique consists of probing the soil with its long bill and smelling out the worm. Betsy G. Bang of Johns Hopkins University finds that relative to the size of the forebrain the olfactory bulb of the kiwi is the second-largest among all the birds she has studied. In the kiwi the ratio of the diameter of the olfactory bulb to the diameter of the cerebral hemisphere is 34 percent. (The snowy petrel shows the highest ratio: 37 percent.) Such evidence with regard to function is only circumstantial, but it finds support in experiments conducted by Bernice M. Wenzel of the University of California at Los Angeles. She gave kiwis access to mesh-covered pots filled with earth, at the bottom of which were ves-

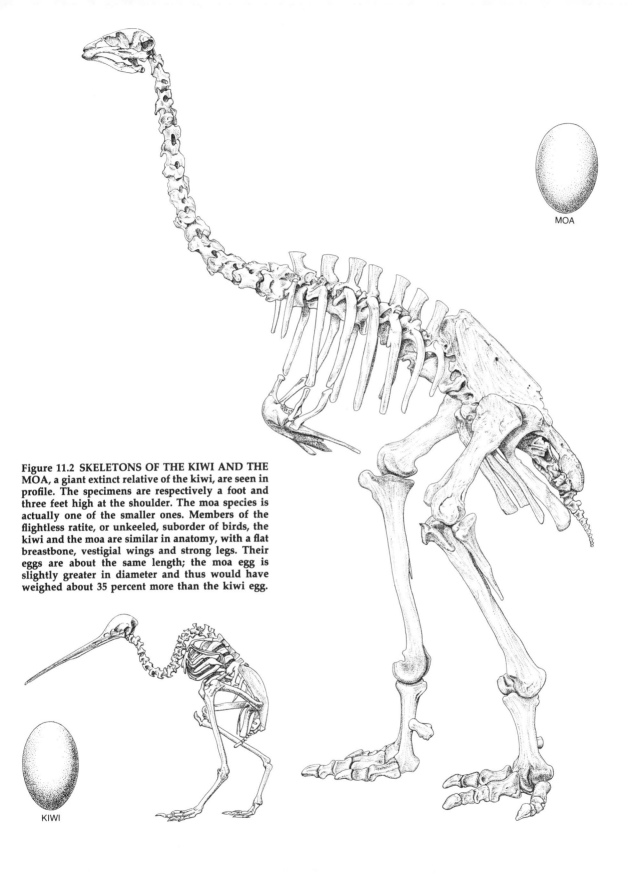

Figure 11.2 SKELETONS OF THE KIWI AND THE MOA, a giant extinct relative of the kiwi, are seen in profile. The specimens are respectively a foot and three feet high at the shoulder. The moa species is actually one of the smaller ones. Members of the flightless ratite, or unkeeled, suborder of birds, the kiwi and the moa are similar in anatomy, with a flat breastbone, vestigial wings and strong legs. Their eggs are about the same length; the moa egg is slightly greater in diameter and thus would have weighed about 35 percent more than the kiwi egg.

MOA

KIWI

sels containing either food or plain earth. If the vessel contained only earth, the kiwis ignored the pot and did not attempt to probe it.

The biology of the kiwi is best appreciated in a context of seemingly diverse matters: the natural environment of New Zealand, the overall evolution of ratite birds and the function of bird eggs and their shells. The topography of New Zealand is spectacularly steep, raised high by faulting and volcanism and sculptured by glacial ice. Rainfall is generally abundant, and before the European colonists began to alter the landscape it was predominantly forest. At the lower elevations and latitudes the forests were a mixture of broadleaf trees and podocarp conifers, a group of trees peculiar to the southwest Pacific and Chile. Above this belt of mixed forest and on up to the timber line, at 900 to 1,500 meters, the forest was predominantly southern beech. Patches of natural grassland were also present, mainly on the drier leeward side of the major mountain ranges such as the area east of the Southern Alps.

Except for two species of bat there were no mammals in New Zealand until man arrived in the first millennium A.D. Hence the evolution of the two ratite families—the kiwis and the moas—could proceed free from the predation and competition that were the rule where terrestrial mammals flourished. It was here that the ratites had their greatest success in diversification. W. R. B. Oliver of the Canterbury Museum in Christchurch, a student of the recently extinct moas, assigned their semifossilized remains to 22 different species. From a recent reanalysis Joel Cracraft of the University of Illinois Medical Center concludes that the fossils represent only 13 species. Nevertheless, combined with the three living kiwi species this is a total of 16 ratite species that are distributed over an area only the size of Colorado, although spanning a range of latitude equal to that between Daytona Beach, Fla., and Goose Bay, Newfoundland.

Some of the moas that evolved in island isolation were only a half meter tall at maturity; others grew to a height of three meters. Dean Amadon of the American Museum of Natural History estimates that the two races of the largest species, *Dinornis giganteus*, weighed between 230 and 240 kilograms, more than twice the weight of an ostrich. In contrast an adult brown kiwi stands no taller than a domestic chicken and weighs on the average 2.2 kilograms.

The first mammals (other than bats) to invade New Zealand were men: a group of seafaring Polynesians, known today only as the moa hunters, who reached the islands in the ninth century A.D. Whether they brought along dogs and rats is uncertain. If they did not, later Polynesian arrivals did.

As the largest land animals in New Zealand until the arrival of man, the moas were obvious prey for the invaders. Although some moa species may have disappeared even before the first human beings arrived, those that survived were probably wiped out before the time of the first European contact: Captain Cook's visit in 1769. The main mammalian invasion began early in the 19th century, when European colonists introduced 22 different kinds of mammals, including pigs, sheep and cattle. Many exotic immigrants were intentionally introduced as game animals or escaped from domestication, and several of them soon began to wreak ecological havoc.

From the time of their arrival the Europeans have put pressure on the islands' long-isolated flora and fauna in a number of ways. The forest habitat of the kiwi shrank under a dual assault: the logging of native trees and the clearing of land for sheep pasture. In recent years much forest land has been cleared and replanted with the imported Monterey pine, because these trees grow faster than the native species. Traps and poisons have been deployed in an attempt to control the imported mammals that became pests. From what is now known about the persistence of poisons in food chains elsewhere in the world it seems likely that the poisons have adversely affected native animals, the kiwi included. Indeed, F. C. Kinsky of the Dominion Museum in Wellington has reported instances of kiwis killed by poisoning. The kiwi is also threatened by trapping, automobiles, predation by dogs and land clearance by burning.

The kiwis are now protected, but no one is sure why the little spotted kiwi is disappearing. Brian Reid and Gordon Williams of the New Zealand Wildlife Service believe that the bird had vanished from North Island before the European settlers arrived. It does not seem likely that it was hunted to extinction by the Polynesians, because the brown kiwi, which is equally vulnerable to hunting, is still present on North Island. To make the puzzle deeper, the process appears to be a continuing one. Reid reports that even though the little spotted kiwi has not been eliminated on South Island, it is rarely seen there.

The closely related moas and kiwis have more distant kin all around the world. In addition to the ostrich of Africa and the emu of Australia, the ratites include the cassowaries of Australia and New Guinea, the rheas of South America and the extinct elephant birds of Malagasy and the Canary Islands. How did these flightless birds evolve?

The anatomy of the ratites makes it clear that their ancestors were flying birds. For example, the ratites' wrist bones are fused for strength at the point where primary flight feathers would be attached, even though their wings are reduced in size and function. Flight entails a high oxygen demand, strong mechanical support and precise coordination, and the physiology of the ratites still shows that their ancestors met these requirements. The birds have retained the air-sac system for effective lung ventilation, a fusion of the lower vertebrae with the pelvis and a shortening of the tailbone for strength, and a large cerebellum for coordination.

All birds are thought to have had a common pigeon-sized ancestor that arose some time in the Jurassic, 150 million years ago, and so it is reasonable to wonder just what factors favored the natural selection of birds as large as the ratites. The advantages of a large body include increased running speed, a larger home range and the potential for dominance over other animals. Another advantage is an improved endurance at times when food is in short supply; this is because the amount of food energy an animal can store increases in linear proportion to its mass whereas the metabolic rate at which the stored energy is burned does not increase.

At the same time a larger body means a greater total food requirement, which is potentially disadvantageous. Perhaps an even more serious disadvantage is that a large bird can weigh too much to fly. C. J. Pennycuick of University College in Nairobi has pointed out that the power required for a bird to become airborne increases out of proportion to the power available from the pectoral muscles. A bird that weighs more than about 12 kilograms is grounded. Thus the giant birds have made an evolutionary trade-off.

Another evolutionary puzzle was a good deal more tangled until recently. This is the problem of how the flightless ratites became so widespread geographically. The ratites were once regarded as primitive birds both because they are flightless and because they are separated from one another by oceans they could not cross. It seemed necessary in those days to construct hypotheses that explained the independent evolution of each member of the group. The biologist T. H. Huxley was almost alone in opposing such hypotheses of polyphyletic origin; he based his opposition on studies of the birds' bony palate.

Today anatomical studies by Cracraft and by Walter J. Bock of Columbia University (relating respectively to the birds' postcranial skeleton and their skull) indicate that the ratites belong among the advanced birds rather than the primitive ones. Chief among their distinguishing features is a palate similar in morphology to that of the tinamou, a partridgelike bird native to South Africa. It may therefore be assumed that the tinamous and the ratites share a common ancestry and that the ratites' departure from the family tree of the flying birds did not precede the rise of the tinamous (see Figure 11.3).

Certain properties of the proteins in the egg white of ratite birds and their immunological relations have been analyzed by David Osuga, Robert E. Feeney, Ellen Prager and Allan C. Wilson at the University of California at Davis and the University of California at Berkeley. Their findings make it possible to construct a ratite family tree very much like one that has been proposed by Cracraft. The only significant difference between the two has to do with the fact that Cracraft relates the ostriches closely to the rheas on the basis of anatomical characteristics. Others have pointed out that the two birds share lice of the same genus and that whereas the ratites of the Pacific have only vestigial wings, the ostrich and the rhea still have wings large enough to serve in courtship display. The immunological studies, however, suggest that the rheas are actually closer to the kiwi than to the ostrich. These findings, regardless of fine points still to be settled, seem to have scuttled the hypotheses of polyphyletic origin, assuming, of course, that the curious geographical distribution of the ratites can be otherwise explained.

Cracraft has noted that a simple explanation is in fact provided by continental drift. The pattern corresponds neatly with the breakup of the great southern continent of Gondwanaland beginning late in the Jurassic period (see Figure 11.4). By the end of the Cretaceous, about 90 million years ago, Africa was completely severed from South America, thus providing the possibility of an early isolation of the rheas and the ostriches that fits the immunological distance between the two. New Zealand began

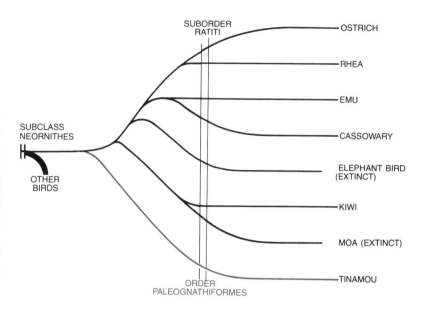

Figure 11.3 FAMILY TREE OF THE RATITES, based on anatomical studies conducted by Cracraft, places the seven families on a common stem with the tinamous: South American birds resembling partridges that have retained the power of flight (*color*). Cracraft considers ostriches and rheas to be the most specialized ratites and closely related. Immunological studies suggest, however, that rheas are more closely related to the kiwis than they appear to be in this diagram.

Figure 11.4 PRESENT DISTRIBUTION OF THE RATITE BIRDS can be explained as the result of a common ancestor having lost the power of flight at a time before continental drift divided the ancient continent of Gondwanaland into the southern continents of today. According to Joel Cracraft, the kiwis and the moas are anatomically more primitive than the other ratites. Their isolation can be attributed to the separation of New Zealand, as is shown here, from western Antarctica 80 million years ago. The elephant birds of Malagasy arose from the ancestral stock left behind, as did cassowaries and emus of Australia and New Guinea, ostriches of Africa and Eurasia and the rheas of South America.

to drift away from western Antarctica about 80 million years ago, leading to the isolation of whatever ancestral ratite gave rise to the moas and the kiwis.

When did the extremes in size among the ratites of New Zealand arise? Is the kiwi perhaps a shrunken moa? The only known moa remains are relatively recent, so that nothing is known of the original ratite stock. If one accepts the monophyletic hypothesis of ratite origin and the likelihood that the birds' ability to fly was lost in the course of their increase in size, one is led to the conclusion that the kiwis are considerably smaller than their ancestors.

Such matters come under the heading of allometry, which Stephen J. Gould of Harvard University has neatly defined as "the study of size and its consequences." For example, if an organism's size is scaled upward, not all of its anatomical features can change in the same ratio if the same function is to be maintained. A skeleton or an eggshell must be disproportionately heavier in the larger organism in order to provide a margin of safety under the stress of greater mass. On the other hand, if the mass of an organism is doubled, its metabolic rate is not doubled but is only increased by 68 percent, and frequencies such as heart rate or breathing rate actually decrease by about 20 percent.

Allometric analysis is gratifyingly simple. This is because within any particular class of organisms not only structural dimensions but also the rates and frequencies of organic processes are exponential functions of body weight. When the functions are plotted on logarithmic scales, they usually show straight-line trends.

The first allometric study of the relation between the mass of bird eggs and the mass of the birds that lay them was published by T. H. Huxley's grandson Julian. For example, a hummingbird weighing 3.5 grams lays an egg weighing .5 gram, and an ostrich weighing 100 kilograms lays an egg weighing between 1.4 and 1.7 kilograms. In other words, although the ostrich egg is far heavier than the hummingbird egg, it is very light in proportion to the weight of the ostrich. Interpolated allometrically, a kiwi the size of a chicken should lay an egg the size of a chicken's or perhaps slightly larger, weighing between 55 and 100 grams. In actuality the egg of the brown kiwi usually weighs between 400 and 435 grams. What should a bird weigh in order to lay a 400-gram egg? The allometric equation answers 12.7 kilograms, about six times the weight of a kiwi hen.

The leading American workers in allometric analysis of bird eggs are Hermann Rahn and his colleagues at the State University of New York at Buffalo: Amos Ar, Charles V. Paganelli, Robert B. Reeves, Douglas Wangensteen and Donald Wilson. Their work has yielded equations that predict from the weight of an egg an amazing range of related properties: the thickness of the shell, the permeability of the shell to gases and water vapor, the probable incubation period, the maximum rate of oxygen uptake and the total water loss during incubation (see Figure 11.5). For example, the diffusion of water vapor from the albumen, or egg white, through the porous shell is measurable as a weight loss over the period of incubation. The rate of loss can characterize the water-vapor conductance (permeability) in the same way that electrical conductivity or resistance is determined when the current and voltage of an electric circuit are known. Oxygen diffuses inward through the same pores, so that the maximum oxygen uptake is directly proportional to the water-vapor conductance. Therefore once one knows the rate of water loss, the metabolism of the embryo can be predicted up to the moment of hatching.

The development of the avian egg begins when the ovum, consisting of a germ cell and a ration of nutrient yolk, is formed in the ovary. The associated albumen, membranes and eggshell are added after ovulation, as the ovum passes down the oviduct. The proportion of yolk to the total mass of the egg is directly related to the extent of embryonic development before hatching (see Figure 11.6). Altricial birds, those that emerge from the egg blind and featherless, come from eggs that are perhaps 20 percent yolk. Precocial birds, which hatch already covered with down and able to feed themselves, come from eggs that are about 35 percent yolk. Given the large size of the kiwi egg and its prolonged incubation (71 to 75 days instead of the 44 days that would be normal for a 400-gram egg) it is no great surprise that the young kiwi hatches fully feathered and receives no food from its parents. What is more noteworthy is the extent of the preparations that lie behind the kiwi's precocity. Reid has found that the contents of the kiwi egg are 61 percent yolk, a proportion half again as large as that found in the eggs of typical precocial birds.

Once the shell of the egg has been deposited no more food or water can be added to the egg. Thus by that time the contents of the egg must include enough yolk to nourish the embryo. They must also

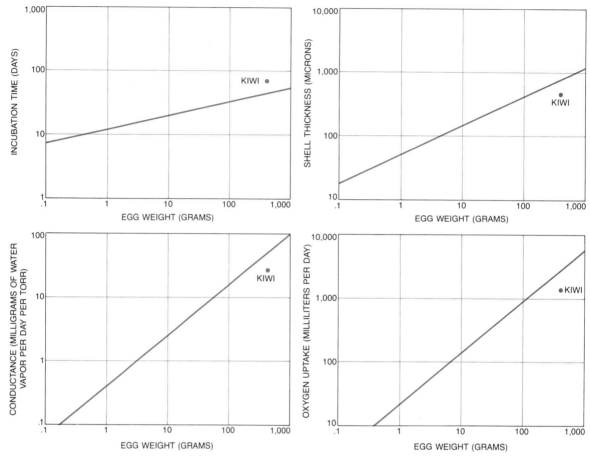

Figure 11.5 RELATIONS AMONG AVIAN EGGS are evident in the straightline plots that appear in these logarithmically scaled graphs plotting increases in egg weight (the horizontal scale) against four factors. These are (top left) increased incubation time, (top right) increased thickness of the eggshell, (bottom left) increased rate of evaporation from the egg and (bottom right) increased oxygen uptake by the egg. In each instance the kiwi egg fails to conform. Its incubation time is some 60 percent longer than the norm for an egg of its size, its shell is nearly 40 percent thinner and its rate of water-vapor diffusion and oxygen uptake are below the norm even though shorter pores of a thinner shell should accelerate diffusion.

include enough water, and the structural integrity of the eggshell must be such that the water is adequately conserved. Rudolph Drent, who is now at the University of Groningen in the Netherlands, has measured the daily rate of water loss from large and small eggs in the course of natural incubation. From Drent's findings, taking into account the relation between incubation time and egg size, Rahn and Ar have established that the percentage of water loss is much the same for all eggs: 18 percent for the smallest eggs and 13 percent for the largest (see Figure 11.7).

It is exciting to see such generalities emerge. Correlations as reliable as these strongly suggest that natural selection operates within quite strict physical limitations that among birds impose a certain order on egg size, incubation time and such structural details as the porosity of the shell. If the evolution of the avian egg has indeed proceeded within, so to speak, close engineering tolerances, what

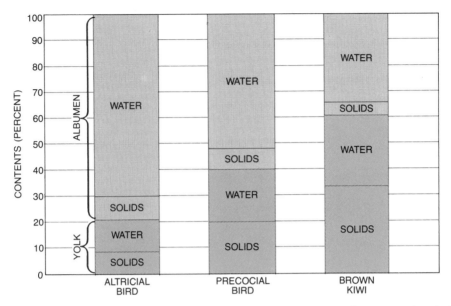

Figure 11.6 CONTENTS OF THREE EGGS are compared. For both the egg and the albumen the solid content is distinguished from the water content. A chick's state of development at hatching is correlated with the percentage of yolk in the egg. The chick of an altricial bird hatches naked, blind and helpless from an egg that is perhaps 20 percent yolk. The chick of a precocial bird, incubated in an egg that is 40 percent yolk, can run after the hen an hour after it has been hatched. The brown kiwi chick receives no parental care. It is sustained in its long incubation by a yolk that represents 61 percent of the egg's contents; the yolk solids are about a third protein and two-thirds fat. The albumen solids are largely protein.

about the kiwi? Are the rules of evolutionary scaling actually understood? Can the rules be real if some birds are exempt from them? It was with just such questions in mind that I recently took a sabbatical leave to visit New Zealand and study kiwis.

One of the questions turning over in my mind during the 14-hour flight from Los Angeles to Auckland concerned the kiwi's extremely long incubation period. Could the extended incubation be the result of poor physical contact for heat transfer between a relatively small bird and a relatively large egg? I looked forward to monitoring egg temperatures.

I had settled down to my first meal in New Zealand, a tea at a restaurant overlooking Lake Taupo, when my attention was drawn to a television report in progress. Barry Rowe of the Otorohanga Zoological Society was being interviewed about his work with kiwis. I stared in disbelief as Rowe opened the door of an artificial kiwi burrow and removed an egg that he opened into two halves, revealing a small transmitter that broadcast a continuous record

of the egg's temperature. After I had recovered from the shock of learning that another investigator had beaten me to the draw I telephoned Rowe. He invited me to share the facilities and hospitality of the zoological society. I promptly joined Rowe in his work and was soon busy measuring the temperatures of reinforced, water-filled eggs as they were incubated.

Rowe recorded the weight changes in a female kiwi during the interval between successive layings (see Figure 11.8). The accumulation of material to be deposited in the energy-rich yolk occupies the first seven and a half days of the 24-day egg-laying cycle. In terms of energy investment the female during that week must exceed her basal metabolic rate by between 174 and 203 percent. The bird's feeding activity in the wild probably increases dramatically in the same period, but there is no quantitative information about kiwi behavior in the wild to confirm this assumption.

The female kiwi, having invested a great deal of

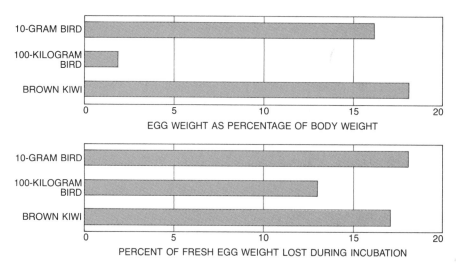

Figure 11.7 WEIGHT OF THE AVIAN EGG is not scaled in linear proportion to the weight of the female. The bars in the top graph show egg weight as a percentage of parental weight for a hypothetical small bird weighing only 10 grams, for a hypothetical large bird weighing 100 kilograms and for a brown kiwi weighing 2.2 kilograms. The weight of the large bird's egg is less than 2 percent of the weight of the parent. The bars in the bottom graph show the weight lost during incubation by the same three eggs, expressed as a percentage of the weight of the fresh-laid egg. The small egg would require incubation for 13.4 days, the large egg for 62.3 days and the kiwi egg for 71 to 74 days. The percentage differences in weight loss are not statistically significant.

energy in egg synthesis over a 24-day period, stops feeding in the last two days of the cycle. As a result when she leaves the burrow after laying her egg she has lost more weight than the weight of the egg alone. The energy investment in the chick-to-be is now completed. The female is free to feed and may have already entered a second cycle of egg synthesis.

The male now takes over the task of incubating the egg for nearly two and a half months. In captivity he leaves the burrow to feed once or twice a night. It has been suggested by some observers that in the wild the female also occasionally brings food to the male during this period: there are reports of males remaining in the burrow several days without feeding.

At embryo depth within the kiwi egg the temperature of incubation averages 35.4 degrees Celsius. When we placed an instrumented egg under a bantam hen, a bird smaller than a kiwi, the temperature of incubation rose to 37.7 degrees. The difference reflects the difference in the body temperature of the two birds. Donald S. Farner of the University of Washington and his colleagues, Norman Chivers and Thane Riney, measured the body temperatures of various kiwi species. They found that the average basal temperature of the adult brown kiwi is 38 degrees C. This is two degrees lower than the body temperature of other birds and more like the temperature of a mammal. In any event if the smaller bantam hen could maintain the kiwi egg at 37.7 degrees, the lengthy period of incubation could not simply be the result of the egg's being too large to be kept warm.

Knowing the natural kiwi incubation temperature, Rowe was able to achieve the first successful artificial incubation of a kiwi egg. Over the 71-day incubation period I measured the oxygen uptake of the egg at intervals of two to three days (see Figure 11.9). Meanwhile, kiwi eggs being difficult to come by, we salvaged infertile eggs from the society's breeding program for other comparative measurements. The kiwi had proved to be a nonconformist not only in egg size but also in incubation temperature and duration. Did the kiwi egg conform in any way to the predictions of Rahn and his associates?

I determined the shells' water-vapor conductance, measured their thickness and separated their yolk and albumen. The separated components were taken to the Ruakura Agricultural Research Centre,

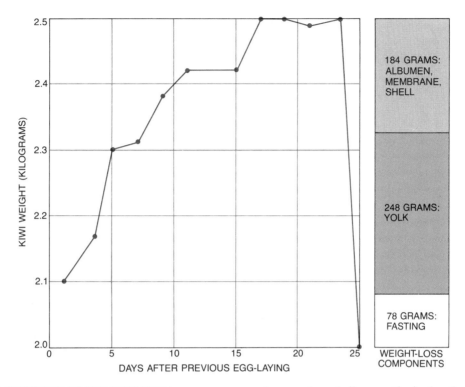

Figure 11.8 SYNTHESIS OF THE KIWI EGG is accompanied by a 400-gram increase in the weight of the female. Some 200 grams are gained during the first seven days of the 24-day cycle. The female's food requirement must be highest during that week. Some 180 grams of albumen, membrane and eggshell are synthesized as the egg travels along the oviduct. Minerals required for eggshell are provided in the diet or are drawn from the bones. For the last two days of the cycle the female fasts. Barry Rowe monitored the changes.

where C. R. Parr, D. P. Karl and R. Whaanga determined the water content of both and prepared samples for measurement of the eggs' energy content by means of the bomb calorimeter. The mean energy content of five eggs was 4.014 kilojoules (959 kilocalories). The figure verified an earlier estimate by Reid, who had assumed a similarity in composition of yolk between the kiwi egg and the eggs of domesticated fowl.

This finding means that in proportion to its size and metabolic rate the kiwi leads all birds in the amount of energy invested in the egg. This enormous energy reserve sustains the kiwi embryo during its lengthy development. Moreover, by the time the kiwi chick emerges from the egg only a little more than half of the yolk has been consumed. What remains in the yolk sac is retracted within the chick's abdomen and provides nourishment for the next two weeks or so.

What about water loss during the 71- to 74-day period of incubation? This period is 61 percent longer than the allometric prediction of the period required to hatch a normal egg of the same weight. With evaporation continuing through the pores for this long a period, one might expect to record a water loss that was 61 percent greater, other things being equal. Other things, we found, were not equal.

The kiwi eggshell is much reduced in porosity compared to the avian average (see Figure 11.10). The reduction results from a decrease in the number of pores and in the size of the individual pores. Consequently the water-vapor permeability of the shell is only 60 percent of the predicted normal rate. This slower-than-normal departure of water vapor from the egg perfectly compensates for the longer-than-normal incubation period. The inward diffusion of oxygen would of course be simi-

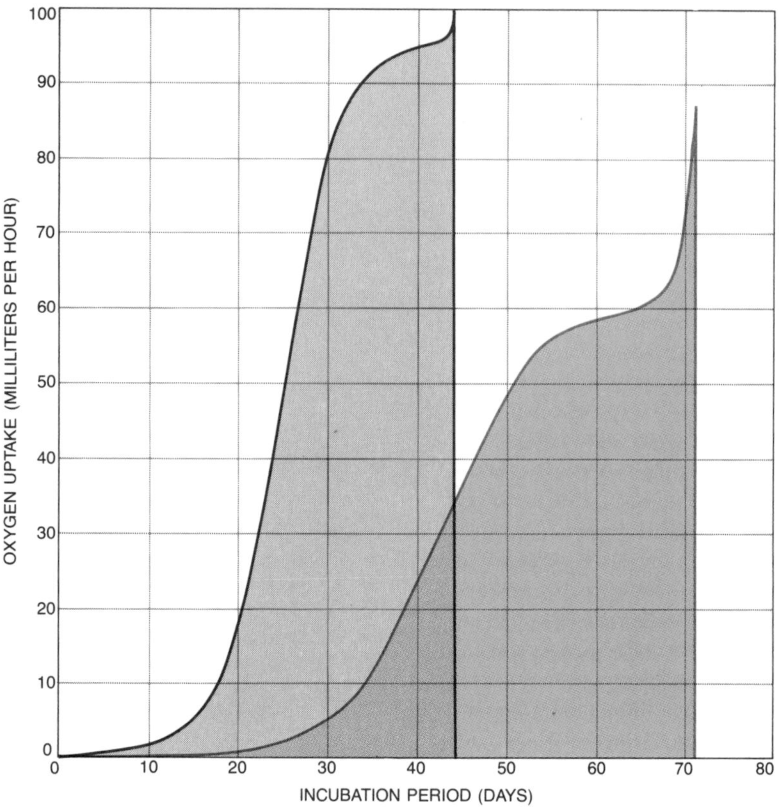

Figure 11.9 UPTAKE OF OXYGEN by the embryo in the kiwi egg over the 71-day incubation period (*color*) is compared with the predicted oxygen uptake for a normal 400-gram egg (*black*). The incubation period for the normal egg is 45 days. As the kiwi embryo passes the first month of incubation the oxygen uptake rises sharply until after the second month it reaches a plateau at about 60 milliliters per hour (1.4 liters per day). Areas under the two curves are the same, showing that the total metabolic cost of development, whether slow or fast, is about the same.

larly slowed, since the same pores are involved. That, in fact, was what my oxygen-uptake measurements indicated: the uptake rate was only 63 percent of normal.

The question arises: Does the embryo's slower oxygen uptake represent a partial oxygen deprivation? Measurements were made of the oxygen consumption of a two-week-old chick that had finally consumed the content of its yolk sac, and also of the oxygen consumption of two older chicks and five adult brown kiwis. For the chicks the mean value for oxygen consumption at rest was 65 percent of the predicted rate for birds of their weight. The adult kiwis averaged 61 percent of the predicted metabolic rate. When individuals of the other two species of kiwis were similarly tested, their metabolic rates also proved to be low. Hence the reduced porosity of the kiwi eggshell limits the oxygen uptake during incubation to roughly the same level that is normal for kiwis after hatching.

The basis of our comparisons was the allometric

equation for the metabolism of all birds except the Passeriformes (the order that includes all songbirds). Could it be that the other ratite birds also have a lower-than-average metabolic rate? In collaboration with Terry Dawson of the University of New South Wales we measured the metabolic rate of emus. They too were found to have a lower rate of oxygen consumption than would be predicted on the basis of their body weight. Perhaps the basal metabolism of all the ratites should be reexamined. It may prove to be below the avian average, as the songbirds are above the average (and the marsupials are below the average for placental mammals).

Just as laying the egg marks the termination of the female's obligations, the male eventually fulfills his obligation at hatching—unless the female has laid another egg. The kiwi chick now takes nourishment from its yolk sac until it is able to feed itself. This life-support system works very well. For example, a kiwi chick that weighed 325 grams immediately after hatching dropped to 225 grams before starting

Figure 11.10 INTERIOR OF A KIWI EGGSHELL is revealed in this scanning electron micrograph, which shows the structure at a magnification of 325 diameters. The crystals of calcite that are secreted by the eggshell gland form a series of roughly circular deposits: the mammillary cones. Continued secretion causes each cone to grow; gaps that remain between cones are respiratory pores of the eggshell.

to gain weight again as a result of foraging for itself. The 100-gram difference in weight was the lunch mama had packed almost three months ago.

Does the ratio of egg weight to adult weight in the kiwi represent an evolutionary decrease in body size or an evolutionary increase in egg size? As one way of finding an answer, let us suppose that the kiwis' ancestors were flying birds that arrived in New Zealand independently of moas or other ratite birds. Then the large egg of the living kiwis, rather than being the legacy of some moa-related ancestor, would have developed by natural selection after the arrival of these hypothetical flying prekiwis.

What are the selective advantages of a large egg? It would favor a more extended embryonic development than is typical of precocial birds, which might serve to schedule energy demands in accordance with the natural abundance of food. The food supply in New Zealand shows little annual fluctuation, however, and so there is no apparent necessity to store up any great reserve of nutrients in the egg.

Several species of birds did reach New Zealand on the wing and have since lost or almost lost the power of flight. Among these are the wekas, two species of now-flightless rails (*Gallirallus australis* and *G. hectori*); the takahe, a flightless water bird of the gallinule group (*Notornis hochstetteri*); the Auckland Island flightless duck (*Anas aucklandica*), and the kakapo, a parrot that can no longer fly but can still glide (*Strigops habroptilus*). Each of these birds, although having evolved to a flightless state, has maintained the same proportion of egg size to body size as that of their flying relatives. This suggests that there is no environmental imperative in New Zealand that equates flightlessness with enlarged egg size. For that matter, moa eggs, although large, were not disproportionately so.

For egg size to increase there must be a selective advantage that overrides whatever forces preserve the general allometric norm. If, however, some selective pressure for a smaller body size once existed, say in the evolution of a kiwi from an ancestral moa, the retention of a disproportionately large egg could merely represent the absence of a strong selective pressure for economy in egg size, perhaps abetted by the advantages of hatching a more developed chick.

The pieces of the kiwi story can be put together in more than one way. I prefer to look on this curious bird as a classic example of convergent evolution. In this view an avian organism has acquired a remarkable set of characteristics that we generally associate not with birds but with mammals. That the temperate, forested New Zealand archipelago provides good habitats for mammals is indicated by the success of the exotic mammals introduced there. When there were no mammals present to lay claim to the niches in this hospitable environment, birds were free to do so.

The kiwi must still lay eggs; after all, it is a bird. It is nonetheless mammallike in a number of ways. For example, Kinsky has reported that kiwis are unique among birds in retaining both ovaries fully functional, so that the female alternates between ovaries during successive ovulations, as mammals do. Also as with mammals the prolonged development of the kiwi embryo proceeds at a temperature below the avian norm. The 70-to-74 day incubation period of the kiwi is much closer to the 80-day pregnancy of a mammal of the same weight than it is to the 44-day period that should be enough to hatch a kiwi-sized egg.

When one adds to this list the kiwi's burrow habitat, its furlike body feathers and its nocturnal foraging, highly dependent on its sense of smell, the evidence for convergence seems overpowering. Only half jokingly I would add to the list the kiwi's aggressive behavior. In the course of my research at the Otorohanga Zoological Society I often had to enter a large pen that was the territory of a breeding male kiwi. When I intruded on his domain at night, he would run up to me snarling like a fighting cat, seize my sock in his bill and drive his claws repeatedly into my ankles until I went away. For this behavior and for the many other reasons I have cited I award this remarkable bird the status of an honorary mammal.

Archaeopteryx

*Although sometimes misclassified or even derided as a fraud,
the prehistoric flier* Archaeopteryx *remains a rich source
of information about the evolution of flight in birds.*

. . .

Peter Wellnhofer
May, 1990

With its reptilian body and tail yet undeniably birdlike wings and feathers, *Archaeopteryx* provides paleontologists with their most conclusive evidence for the evolution of birds from reptiles. This pigeon-size prehistoric bird is known from only six fossil skeletons and the imprint of one lone feather, but paleontologists have deducted a wealth of information from those few specimens. From the time of its discovery more than a century ago, *Archaeopteryx* has been the object of heated discussions between critics and defenders of Charles Darwin's theory of evolution. *Archaeopteryx* has nonetheless defended its reputation and evolutionary theory against all challenges —even recent allegations of fraud.

In 1985 the British astronomer Fred Hoyle charged that the *Archaeopteryx* specimen in the British Museum of Natural History was a fake. He claimed that a forger had created the specimen by first applying a thin layer of binding material mixed with pulverized rock to the fossilized skeleton of *Compsognathus*—a type of small dinosaur called a theropod—and then making impressions of feathers in it. Hoyle and his colleagues also suggested that the other *Archaeopteryx* fossils either were forgeries or did not really show imprints of feathers. In England, *Archaeopteryx* soon became known as the Piltdown chicken.

Because of the publicity surrounding the affair, the British Museum decided in 1987 to stage a special exhibition to accompany the scientific reexamination of its fossil. Various tests proved that the stone in which the feather imprints were found did not differ in structure or composition from the surrounding material. Notwithstanding the modern appearance of *Archaeopteryx*'s feathers, nothing pointed to their being forgeries. Moreover, the stone plates encasing the skeleton fit together perfectly, which would have been impossible if the fossil had been tampered with by adding a layer of cement.

Ironically, the features that Hoyle saw as proof of the fossil's inauthenticity—its mixture of *Compsognathus*-like bones and modern feathers—are some of the most important clues that paleontologists have for understanding how birds and bird flight evolved. Its combination of anatomical characteristics from two distinct classes of animals make *Archaeopteryx*, the oldest-known bird, a textbook example of a transitional form between reptiles and modern birds.

The original discovery of *Archaeopteryx* came as if custom-ordered by the Darwinists. In 1861, only two years after the publication of Darwin's *The Origin of Species by Means of Natural Selection*, a fossilized skeleton with imprints of feathers was discovered in the limestone quarry at Solnhofen in Bavaria and passed into the possession of Carl Häberlein of Pappenheim, who later sold it to the British Museum.

This fossil, usually called the London specimen, was not the first evidence that birds existed in the late Jurassic period, 150 million years ago. One year before the discovery of the London skeleton, a worker in the Solnhofen quarry had found the impression of a single feather (see Figure 12.1). Until then the oldest bird fossils were known only from the Tertiary period, almost 100 million years after the Solnhofen limestone formed.

In 1861 Hermann von Meyer, a paleontologist at the Senckenberg Natural History Museum and Research Institute in Frankfurt, reported that the feather imprint "is a real fossil and that it matches the feather of a bird perfectly." In the same report he also mentioned the London specimen: "[T]he almost complete skeleton of an animal covered with feathers had been found in the lithographic slate. It is said to show many deviations from the modern birds. I will publish the feather that I investigated with an exact illustration. I deem the name *Archaeopteryx lithographica* appropriate for the animal."

With this, the scientific name for the bird from Solnhofen, *Archaeopteryx lithographica*, was introduced. The name *Archaeopteryx* means "old wing," and *lithographica* is a reminder that limestone from Solnhofen was usually called lithographic slate during the 19th century. Only the stone from the Solnhofen quarry was sufficiently hard, compact and fine-grained to be used in the lithographic printing process. These same qualities preserved the bones and imprints of the feathers of *Archaeopteryx* with incredible delineation and clarity.

The geologic conditions that led to the formation of the Solnhofen limestone explain the exceptional preservation of the *Archaeopteryx* fossils. During the latter part of the Jurassic period, the area of what is today the southern Franconian Alb was a tropical lagoon divided into various basins by submarine reef complexes. North of this lagoon was the landmass of what is now central Germany; south of it was the Tethys Sea.

Figure 12.1 SINGLE FEATHER (*left*) from *Archaeopteryx*, which was unearthed in 1860, is virtually identical to the feathers of modern birds. These similarities extend even to the microscopic level, as shown in a closeup of tail feathers from the Berlin specimen (*right*), enlarged by about 500 percent. Detailed fossils of *Archaeopteryx* feathers have survived because of the hardness and fine grain of the Solnhofen limestone.

The region was not a South Seas paradise: the water in the lagoon was too salty and contained almost no oxygen. It was therefore inhospitable to most forms of life. Occasional storms caused floods that flowed over the southern reef barriers and carried multitudes of marine animals and plants into the basins. These died quickly in the lagoon waters, sank to the bottom and were promptly buried by layers of lime-rich mud.

Plants and animals from the northern landmass and its outlying islands also reach the lagoon at Solnhofen. These lands were home to many types of life: conifers, cycad ferns, ginkgo trees, insects, dinosaurs and *Archaeopteryx*. Tropical storms could have brought flying creatures to the lagoon by blowing them out to sea; currents could also have carried plants and animal carcasses there. Because almost no carrion feeders or microorganisms lived in the salty lagoon, dead organisms decayed little before fossilizing in the lime-rich sediments.

Six *Archaeopteryx* skeletons have been found so far, all in the Solnhofen limestone dating to the late Jurassic period (see Figure 12.2). They are the oldest-known bird fossils. (Sankar Chatterjee of Texas Technical University in Lubbock has identified parts of fossil skeletons from much older Triassic strata in Texas as those of a bid that he calls *Protoavis*, but those skeletons are fragmentary and evidence for their avian nature has not yet been presented.) Each of these specimens contributes to the current understanding of the behavior and morphology of *Archaeopteryx* and of the origin of birds.

The London specimen represents almost the entire skeleton of *Archaeopteryx*. The skull, however, exists in fragments, and only the structure enclosing

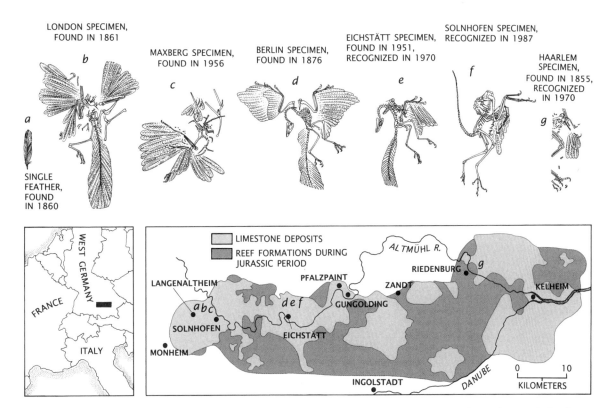

Figure 12.2 SIX SKELETONS and one isolated feather imprint are the only known fossils of *Archaeopteryx*. All were recovered from a region in the southern Franconian Alb in West Germany (*left*), which was a tropical lagoon subdivided by reef formations during the Jurassic period. The letters on the map (*a–g*) show where each fossil was found. Some of the fossils were initially classified as pteranodons or other dinosaurs and were not recognized as *Archaeopteryx* until years after their discovery.

the brain and parts of the outer toothed jawbones have been preserved. In addition to the clear impressions of plumage on the wings and on the tail, another typical avian feature is present: the furcula, or wishbone, which evolved through fusion of the collarbones. Until only a few years ago the furcula was thought to be unique to birds; however, furculas have since been found in some dinosaurs from the Cretaceous period.

The second specimen was discovered in the fall of 1876 in a quarry near Eichstätt and later sold to Ernst Häberlein—the son of the man who called paleontologists' attention to the first specimen. He had initially categorized the fossil as that of a flying reptile. Only after Häberlein removed the overlying layers of stone did the well-preserved imprints of feathers come to light, prompting the reclassification of the fossil. The Museum for Natural History of Humboldt University in Berlin eventually acquired this *Archaeopteryx* specimen in 1881.

Known as the Berlin specimen, it is better preserved than the one in London (see Figure 12.3). The articulated skeleton is in a natural position, which means that at the time of the animal's entombment at the bottom of the Solnhofen lagoon, decomposition had not yet begun. The skull has teeth like those of a reptile. The extreme backward bend in the neck, caused by the pull of the ligaments after the muscles relaxed, is a peculiarity of dead birds but has also been observed in fossils of flying reptiles and some other small dinosaurs with long necks, such as *Compsognathus*.

The wings of the Berlin specimen show the imprints of well-preserved feathers. The three "fingers" of the wing (or manus, as it is usually called by anatomists) were evidently movable and equipped with sharp, strongly bent claws. The fingers of modern birds are shorter, partially fused together and clawless.

The long, saurianlike tail of the Berlin specimen displays a biserial plumage; that is, there are two sets of tail feathers, each arranged symmetrically in a horizontal plane along the column of the tail vertebrae. The imprints of the feathers are especially rich in details because the entire structure of the feathers, down to the interlocking barbs, is visible.

After the discovery of the Berlin specimen, 80 years passed before another one was found. In a quarry not far from the site of the discovery of the London specimen, the fossil remnants of another winged animal were spotted in 1956. After analyzing them, Florian Heller, a paleontologist from Er-

langen University, concluded that the specimen matched the one in London and was therefore *Archaeopteryx lithographica*. The fossil, which is privately owned, was lent to the Maxberg Museum near Solnhofen until 1974, which is why it is commonly known as the Maxberg specimen.

The Maxberg *Archaeopteryx* must have floated on the water for a long time after its death because its head and tail are missing: they must have separated from the body before fossilization occurred. The hind legs and the wings had detached from their natural positions but—judging from the orientations of the feathers—were still held together by tendons.

T he fourth specimen of *Archaeopteryx* came to light after lying unrecognized in the inventory of the Teylers Museum in Haarlem, the Netherlands, for over a century. It was unearthed from a quarry in 1855 and was therefore actually found before the London specimen. Yet the specimen was misclassified as a pterodactyl as early as 1857. Not until 1970 did John H. Ostrom of Yale University recognize that its skeletal characteristics were typical of *Archaeopteryx*. The specimen was poorly preserved: all that remain are parts of bones from the left lower arm and wing, the pelvis and the hind legs. The claws on the wings and the feet are nonetheless remarkably well preserved.

The fifth specimen, too, was misidentified at first. It was found in a quarry in the area of Eichstätt in 1951, five years before the discovery of the Maxberg specimen. This skeleton is smaller than any of the others but is nearly complete; it was initially taken to be a small reptile similar to the chicken-size *Compsognathus*. Not until 1970 was it recognized as an *Archaeopteryx* by Franz X. Mayr of the University of Eichstätt, when he illuminated the fossil from the side to reveal the faint impressions of wing and tail plumage.

This Eichstätt specimen of *Archaeopteryx* has the best-preserved skull of any of the fossils (see Figure 12.4). Recent computerized tomographic scans have shown that the articulation of the quadrate bone

Figure 12.3 BERLIN SPECIMEN of *Archaeopteryx* was found in 1876 in a quarry near Eichstätt. The skeleton is in a natural position, which means when the animal was entombed in the Solnhofen lagoon, decomposition had not yet begun. Therefore, this specimen is one of the best preserved of the six *Archaeopteryx* skeletons that have been found.

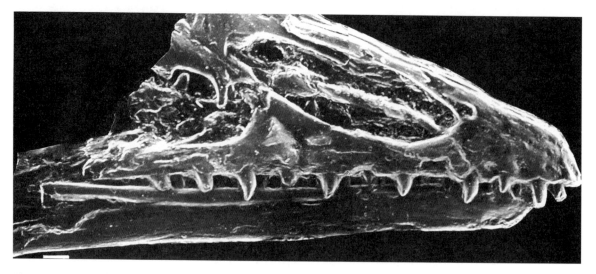

Figure 12.4 SNOUT of the small *Archaeopteryx* specimen in Eichstätt, West Germany, has been enlarged 6.5 times in this electron micrograph. Because of its backward-curved teeth, the fossil had once been classified as *Jurapteryx* **recurva*. Analyses of the bones now make it seem more likely that this specimen is an *Archaeopteryx lithographica* that had died before it matured.**

with the brain case is almost identical to that in modern birds. Based on the backward-bent neck and the good preservation and position—identical to those of the Berlin specimen—one can surmise that both animals died in similar ways.

Certainly the Eichstätt bird did not die of old age: the small skeleton suggests that the animal was only a juvenile. The metatarsal bones in the feet show no sign of fusion as those of the larger Maxberg specimen do. Also, the furcula is missing from the otherwise complete skeleton. The best explanation is that the immature furcula was still cartilaginous, not bony, at the time of death and consequently was not preserved in the fossil.

Another feature of the small Eichstätt *Archaeopteryx* is its relatively long hind legs. Their length indicates that the leg bones matured earlier than the wings and some other body parts. Young animals may have needed their hind legs for walking because their ability to fly may have developed only later in life.

The sixth and most recently discovered specimen of *Archaeopteryx* came to the world's attention in 1987, when Günter Viohl, the curator of the Jura Museum in Eichstätt, spotted the prehistoric bird in a collection of fossils belonging to Freidrich Müller,

a former mayor of Solnhofen. No imprints of feathers were apparent, and most of the skull had been lost. Because of its long, strong hind legs and long tail, the fossil was initially mistaken for that of a *Compsognathus*. The fossil, which is on display in the Bürgermeister Müller Museum, now belongs to the village of Solnhofen.

All the recovered body parts in the Solnhofen specimen are in their natural, articulated position. Under low-angle illumination from the side, the left wing reveals small, curved impressions of shafts from the main feathers; the outer boundary of this wing is also well marked by impressions. There are no such signs on the right wing and the tail, but this can be explained by the position of the skeleton: the carcass of the prehistoric bird had sunk to the bottom of the Solnhofen lagoon on its left side, and its left wing had been deeply anchored in the protective mud on the bottom. The feathers on the unprotected parts of the body could easily have been swept away by currents.

What is immediately striking about the Solnhofen specimen is its size. On the basis of the length of the wings, it is 10 percent larger than the London specimen—the largest previously known—and 50 percent larger than the small Eichstätt specimen. This *Archaeopteryx* was fully the size of a modern chicken.

Before one can make sense of the six fossil skeletons (and the single feather imprint), it is essential to address the important question of whether they are, in fact, all fossils of the same species. Indeed, their classification has always been controversial. Over the years many names have been given to the various specimens by paleontologists attempting to sort them into different species and even different genera.

The biological definition for a species—a group of actually or potentially crossbreeding populations—is useless for a paleontologist, who cannot test long-dead specimens by that criterion. Paleontologists usually have no choice but to define ancient species by their skeletal morphology. Working from incomplete information, paleontologists attempt to discriminate between the differing characteristics of species and the variations attributable to age, sex and other individual features. A species defined by paleontological deductions and one defined by biological traits are not necessarily identical.

Part of the problem with classifying *Archaeopteryx* stems from ignorance of whether their growth pattern, as recorded in their bones, was more reptilian than birdlike. Reptiles grow throughout their lives (although the rate slows during advanced age); conversely, birds quickly attain a characteristic adult size. In reptiles the centers of growth are in the shafts of their hollow bones, whereas growth in young birds takes place at the bones' thick cartilaginous ends, called epiphyses. During the final stage of a bird's growth, its epiphyses turn from cartilage into bone, leaving a scar that disappears when the bird matures.

None of these *Archaeopteryx* specimens shows any such scars on its hollow bones. If the growth of these animals was birdlike, then the various specimens might indeed represent different species. On the other hand, if they had reptilian growth patterns—an assumption supported by the predominantly reptilian traits of the skeletons—then the specimens could clearly be members of the same species, differing in size and age. Recent studies by Marilyn A. Houck and Richard E. Strauss of the University of Arizona and Jacques A. Gauthier of the California Academy of Sciences support the view that the six specimens of *Archaeopteryx* represent different stages of growth of a single species.

These skeletal considerations and a complex set of other unknowns—such as whether the specimens lived hundreds of thousands of years apart or whether they differed in sex—lead me to conclude that the fossils should be assigned to one pragmatic classification: *Archaeopteryx lithographica*.

The birdlike appearance of *Archaeopteryx* raises the obvious question of whether or not the animal could fly. It is pertinent that no sternum, or breastbone, can be found in any of the specimens, even the large, presumably mature Solnhofen animal. Apparently, *Archaeopteryx* had not developed an ossified (or bony) sternum, a structure important to flight in today's birds.

The sternum of modern birds is a wide, arched shelf of bone that often extends from the chest to the pelvic area and serves as a protective, supportive bowl for the internal organs during flight. In the middle of the outer side of the sternum is a crest that acts as an anchoring point for the pectoral muscles. Compared with the rest of the body, the size of modern birds' pectoral muscles is unmatched by any other animal; these enormous muscles enable birds to fly by flapping their wings.

There are no indications that *Archaeopteryx* had similarly developed pectoral muscles. Instead of a sternum it had gastral (abdominal) ribs, just as its saurian ancestors did (see Figure 12.5). Gastral ribs are thin, fishbonelike braces in the abdominal area that are not fixed to the rest of the skeleton. They are found today in lizards and crocodiles and were relatively common in early reptiles and amphibious animals. The gastral ribs may have protected the abdominal area and helped to support the internal organs of *Archaeopteryx*, but they could not have served as points of attachment for the pectoral muscles.

Yet *Archaeopteryx* did have a furcula like that of today's birds. In modern birds some of the pectoral muscles attach to that structure; *Archaeopteryx* may therefore also have had a small area of attachment for those muscles on its furcula. Its ability to fly, however, would have been rather limited.

There are other indications that the prehistoric bird was not a good flier. In modern birds, air bags extend from the lungs into the body and reach into the bones through small openings that are usually found at the top end of the upper-arm bone. These air bags enhance the capacity to breathe and help the bird to meet its heavy oxygen requirements during flight. *Archaeopteryx* lacks openings for air bags in its bones; therefore, it is questionable whether the animal had birdlike lungs.

Also, the bones in the manus of *Archaeopteryx* are not fused to support the wing as they are in modern

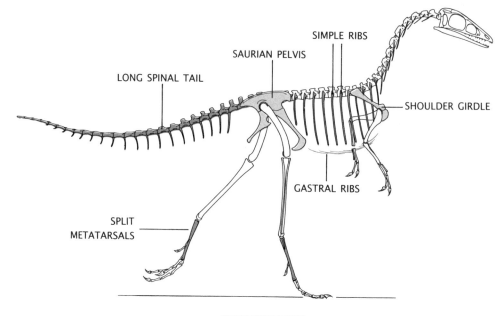

LONG SPINAL TAIL

SAURIAN PELVIS

SIMPLE RIBS

SHOULDER GIRDLE

GASTRAL RIBS

SPLIT METATARSALS

COMPSOGNATHUS

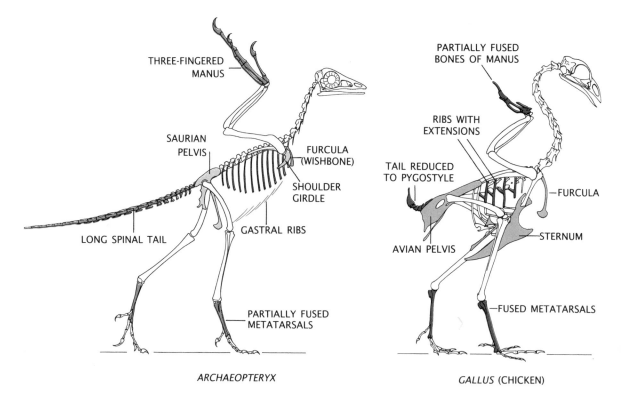

THREE-FINGERED MANUS

SAURIAN PELVIS

FURCULA (WISHBONE)

SHOULDER GIRDLE

GASTRAL RIBS

LONG SPINAL TAIL

PARTIALLY FUSED METATARSALS

ARCHAEOPTERYX

PARTIALLY FUSED BONES OF MANUS

RIBS WITH EXTENSIONS

TAIL REDUCED TO PYGOSTYLE

FURCULA

STERNUM

AVIAN PELVIS

FUSED METATARSALS

GALLUS (CHICKEN)

birds. Its fingers could move independently of one another and were equipped with strong, pointed claws. The largest feathers of the manus originate from only the middle finger; the largest feathers of the arm come from the ulna, the largest bone of the lower arm. Yet the ulna is smooth, in contrast to that of modern birds, which has small knobs where the main feathers are anchored firmly to the bone by ligaments. It therefore seems that the main feathers of *Archaeopteryx* were not anchored in the skeleton.

The underdeveloped pectoral muscles, the reptilian lungs and the lack of firm anchoring for the main feathers all paint a picture of *Archaeopteryx* as a poor flier.

Nevertheless, the perfectly developed plumage of *Archaeopteryx* makes it certain that the animal did fly. No other vertebrates besides birds are equipped with real feathers; feathers must have played a decisive role in the evolution of flight.

It is currently thought that feathers evolved from reptilian scales. Did feathers or featherlike structures originally insulate warm-blooded dinosaurs from the cold? Did they protect cold-blooded reptiles from the heat and sun? Were the feathered arms used to attract mates and to battle with sexual rivals, or did they form a basket for catching insects? All these ideas and others have been proposed, but they remain only theories.

Only one point is certain: *Archaeopteryx* represents an advanced stage in the evolution of flight. Its main feathers show the asymmetric, aerodynamic form typical of modern birds. This similarity proves that the feathers of *Archaeopteryx* must have been used for flying. Their sophisticated form also hints that the hypothetical ancestor of *Archaeopteryx*

must have had feathers of some kind, too, although it probably had not yet acquired the ability to fly.

Although *Archaeopteryx* could not have flown long distances, it was capable of wing-flapping flight and was also a good runner. Indeed, the architecture of the pelvis and hind legs of the prehistoric bird suggest it was adept at moving on the ground. Its pelvis resembles the three-pronged design of the saurischian dinosaurs, especially the bipedal theropods such as *Compsognathus*. The pelvic and leg muscles of *Archaeopteryx*, therefore, must also have been roughly saurian.

Archaeopteryx must have stood on its hind legs much as *Compsognathus* and the other theropods did. The posture of these animals contrasts with that of modern birds, whose body is suspended at the pelvis like a seesaw when the thighbones are almost horizontal. Like *Compsognathus*, *Archaeopteryx* did not topple forward because of the counterweight of its tail, which was about as long as the body.

The tail was flexible near its base but became increasingly inflexible toward its tip because of bony protrusions on the 23 tail vertebrae. Such protrusions are present in some bipedal dinosaurs and in the long-tailed flying saurians of the Triassic and Jurassic periods. The rigidity of the tail helped the animal to balance itself during abrupt changes of direction while running or flying. In addition, the plumage of the tail provided an aerodynamically stabilizing horizontal surface.

In modern birds the tail vertebrae have shrunk and fused into a pygostyle, one of the few structures that are present only in birds. In the birds that immediately succeeded *Archaeopteryx*, the shortening of the vertebral column of the tail must have been accompanied by a gradual shift of the center of gravity toward the front. To compensate in part for this forward shift, the muscles in the pelvic region enlarged, and there was a corresponding increase of surface area on the pelvic bone to which muscle could attach. During the reorganization of the pelvis, the two fused pubic bones turned backward and separated. The job of supporting the internal organs was then taken over by the sternum, which developed simultaneously.

The foot of *Archaeopteryx* is definitely adapted to running and has features intermediate between those of reptiles and modern birds. In reptiles the metatarsal bones in the foot are separate; in modern birds these bones have fused into a single bone. X-ray studies of the Maxberg specimen by Wilhelm

Figure 12.5 ANATOMICAL COMPARISONS between the small dinosaur *Compsognathus*, *Archaeopteryx* and a chicken show that the evolution of flight involved structural changes throughout the body, as essentially avian features (*red*) evolved from reptilian ones (*blue*). *Archaeopteryx* had longer forelimbs than *Compsognathus*, but its three movable fingers were not short and fused together as in modern birds. *Archaeopteryx* did not have a bony sternum, which is essential for strenuous wing-beating, but it did have a furcula (wishbone). Its food and hip structures indicate that *Archaeopteryx* was adept at walking. Its tail had not regressed to become a stubby pygostyle and could have counterbalanced the weight of the body when *Archaeopteryx* walked or ran.

Stürmer, a physicist and paleontologist from the Siemens Company in Erlangen, have revealed that the metatarsals had partially fused. In the biggest specimen, the one from Solnhofen, the bones had fused even more. These observations suggest that in *Archaeopteryx* the metatarsals ossified and fused with age.

Overall, the foot structure of *Archaeopteryx*, like that of its theropod ancestors, is birdlike, with three long toes and a short backward-facing toe. The sharp, bent claw on the backward-facing toe suggests this prehistoric bird might have been able to grasp objects with its feet and perch on a tree branch.

In summary, the evolution of flight was accompanied by the reorganization of more than the structures most needed for flying; the changes encompassed the animal's entire skeleton and physiology.

There are two basic, contradictory models for the evolution of flight. According to the arboreal model, flight involving the downward beating of the wings evolved from gliding and originated in animals that climbed and leaped from trees. The alternative, cursorial model posits that flying arose among bipedal animals making small jumps (to catch insects, for example) while running and flapping their forelegs simultaneously to extend the jumps. With progressive development of the wing structures, the jumps became longer and higher until the animals could eventually keep themselves on a flying course by beating their wings.

Evidence for the cursorial theory can be found in the physical adaptations of *Archaeopteryx* for walking and running. Yet the energy requirements for running with simultaneous wing-beating would be very high, particularly during the early stages of the evolution of flight. Moreover, flying up from the ground entails fighting gravity, whereas gliding down from a tree takes advantage of gravity and therefore requires less energy.

The arboreal model assumes that *Archaeopteryx* and its ancestors were able to climb trees. Do the claws of *Archaeopteryx* support this idea? Its claws are bent in the form of a sharp sickle, with a cutting edge on the inside and reinforcing material on the outside (see Figure 12.6). Similar claws can be found in bats, squirrels and woodpeckers—all animals that climb tree trunks and cling to bark. The claws of predatory birds and animals that run along the ground are quit different. Today's birds climb exclu-

sively with the claws on their feet; *Archaeopteryx* would also have been able to use the claws on its fingers, in particular the one on its flexible first finger, to hook and anchor itself, while the tail provided additional support.

A model that combines features of both the arboreal and cursorial theories might be called the arbocursorial theory or, more descriptively, the climber-runner theory (see Figure 12.7), and is based partly on the ideas of Walter J. Bock of Columbia University.

According to the arbocursorial theory, *Archaeopteryx*'s predecessors were small, probably bipedal reptiles that took to the trees during the late Triassic and early Jurassic periods, about 200 million years ago. Forests may have served as places for hiding, nesting and breeding; they may also have offered advantages in the search for food. The beginning of an arboreal life for these protobirds was probably linked to the evolution of warm-bloodedness and the simultaneous evolution of feathery structures as insulation for maintaining a constantly high body temperature. Life in the trees would also have promoted the development of stereoscopic vision and the ability to orient in three-dimensional space, both of which are prerequisite skills for flying.

Large air-resisting feathers, particularly those on the forelegs, would have softened the landings of these protobirds during leaps to the ground by slowing their descent. Gliding could have evolved out of these slow falls, and the ability to maintain a straight line of flight could have emerged by flapping the wings.

Because the feet of *Archaeopteryx* are adapted for running, the ability to move on the ground must have been important to the prehistoric bird and its protobird ancestors. Gliding between trees and landing on branches require precise steering ability, which in turn requires great coordination. Simple, parachutelike landings on the ground would therefore have been much easier for the earliest birds to control than complicated landings on trees. Once reaching the ground, the animals would have charged toward the next tree and climbed up in search of insects, nesting space or refuge.

Was *Archaeopteryx* the ancestor of all later birds? What can be deduced about avian evolution from the fossils of the succeeding Cretaceous period? The only relatively complete skeletons are

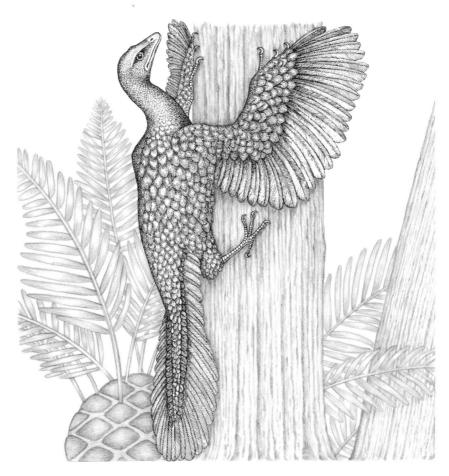

Figure 12.6 CLIMBING with the sharp, bent claws on its fingers and toes, *Archaeopteryx* could have ascended a tree in preparation for a flight or to find food, mates or shelter.

from the latter part of that period (roughly 85 million years ago). Those birds had teeth and were partially adapted to a diving, aquatic way of life. The direct descent of such specialized forms from *Archaeopteryx* scarcely seems possible. Consequently, *Archaeopteryx* has often been regarded as an evolutionary dead end for birds.

Recently, however, bird fossils from the early Cretaceous period (roughly 125 million years ago) have been found that seem to represent an intermediate stage between *Archaeopteryx* and modern birds. In particular, the skeleton of a small bird found in the limestone of Las Hoyas in eastern central Spain in 1984 displays a combination of ances-

tral and modern characteristics. Its pelvis and hindlegs seem more reptilian than those of today's birds; its shoulders and furcula seem more modern than those of *Archaeopteryx*. Its most intriguing characteristic, however, it is pygostyle, which has 15 fused vertebrae. This is longer than the pygostyle of modern birds (which has from four to 10 fused vertebrae) but shorter than the tail of *Archaeopteryx* (with 23 vertebrae).

The bird from Las Hoyas, like *Archaeopteryx* itself, illustrates that the early evolution of birds was strongly influenced by the physical requirements of flight. It is not currently possible to tell whether or not *Archaeopteryx* was the direct ancestor of the bird

Figure 12.7 *ARCHAEOPTERYX* was not a strong flier but could launch itself from trees, according to the arbocursorial theory of the evolution of flight. The prehistoric bird might have made parachutelike landings on the ground; it then would have run to a nearby tree.

from Las Hoyas and of all other birds, but this correlation is not of major importance. What is significant is that the six known skeletons of *Archaeopteryx* and the single feather provide clues to how birds evolved. As Adolf Portmann, a zoologist from the University of Basel, said about the fossils in 1957: "They are documents without which the idea of evolution would not be as powerful."

The Authors

The Editor

DOUGLAS W. MOCK is associate professor of zoology at the University of Oklahoma. He received his B.S. from Cornell University (1969), his M.S. from the University of Minnesota (1972), his Ph.D. from the University of Minnesota (1976) and he was a Postdoctoral Fellow at the Smithsonian Institution (1976–1977). An evolutionary behaviorist, Dr. Mock has written numerous articles and papers on birds and bird behavior, many of them reflecting his pioneering studies of sibling rivalry. He joined the University of Oklahoma in 1978 as an assistant professor and became an associate professor in 1983.

JAMES L. GOULD and PETER MARLER ("Learning by Instinct") are respectively professor of biology at Princeton University and professor of animal behavior at the University of California at Davis. Gould received a B.S. from the California Institute of Technology in 1970 and a Ph.D. from Rockefeller University in 1975, then joined Princeton's faculty. Marler received a B.Sc. from the University of London in 1948 and a Ph.D. in 1952; in 1954 he earned a second Ph.D. from the University of Cambridge. After nine years on the faculty of the University of California at Berkeley, he joined Rockefeller in 1966. He left Rockefeller to join the University of California at Davis in 1988.

EBERHARD GWINNER ("Internal Rhythms in Bird Migration") is head of the Vogelwarte Radolfzell, an ornithological station of the Max Planck Institute for Behavioral Physiology, and an adjunct professor of zoology at the University of Munich. He received his Ph.D. from the University of Tübingen in 1964. Subsequently he worked at the institute, the University of Washington and Stanford University.

SARA J. SHETTLEWORTH ("Memory in Food-hoarding Birds") is professor of psychology at the University of Toronto. She was graduated from Swarth-more College in 1965 with a B.A. and received her M.A. in 1966 from the University of Pennsylvania and her Ph.D. in 1970 from the University of Toronto. She had done work on the role of learning and memory mechanisms in foraging behavior and on sea-turtle behavior with her husband, Nicholas Mrosovsky.

FERNANDO NOTTEBOHM ("From Bird Song to Neurogenesis") is professor of animal behavior at Rockefeller University and director of the university's Field Research Center for Ethology and Ecology. A native of Argentina, he earned a B.A. and Ph.D. in zoology from the University of California, Berkeley. He joined the faculty at Rockefeller as assistant professor in 1967 and was made full professor in 1976.

ERIC I. KNUDSEN ("The Hearing of the Barn Owl") is assistant professor of neurobiology at the Stanford University School of Medicine. He earned his degrees in the University of California system: an A.B. in 1971 and an M.A. in 1973 (both at Santa Barbara) and a Ph.D. in 1976 (from San Diego). He began his research career as an undergraduate working on the bioluminescence of the sea pansy, moved progressively up the phylogenetic ladder, from the horseshoe crab to the catfish to the owl.

GERALD BORGIA ("Sexual Selection in Bowerbirds") is associate professor of zoology at the University of Maryland at College Park. He received a bachelor's degree from the University of California at Berkeley in 1970 and his master's degree (1973) and Ph.D. (1978) from the University of Michigan. After postdoctoral work at the University of Chicago and the University of Melbourne he became assistant professor at the University of Maryland in 1980.

BRUCE M. BEEHLER ("The Birds of Paradise") is a zoologist at the Smithsonian Institution's National Museum of Natural History. He earned his Ph.D. at Princeton University (1983). Beehler has spent some 60 months in the rain forests of New Guinea studying the ecology of the bird fauna and the ecology and biogeography of the Indian avifauna in collaboration with S. Dillon Ripley.

PETER B. STACEY and **WALTER D. KOENIG** ("Cooperative Breeding in the Acorn Woodpecker") are zoologists. Stacey received his B.A. at Middlebury College and his M.A. and Ph.D. from the University of Colorado at Boulder. After two years of postdoctoral study at the University of Chicago he became assistant professor of life sciences at Indiana State University. He is now a professor at the University of Nevada, Reno. Koenig received his Ph.D. from the University of California at Berkeley, then taught at Occidental College for a year. In 1982 he joined the Hastings Natural History Reservation of the Museum of Vertebrate Zoology, as assistant research zoologist and the University of California at Berkeley as associate research zoologist.

J. DAVID LIGON and **SANDRA H. LIGON** ("The Cooperative Breeding Behavior of the Green Woodhoopoe") are biologists. David Ligon is professor of biology at the University of New Mexico. His degrees are from the University of Oklahoma (B.S., 1961), the University of Florida (M.S., 1963) and the University of Michigan (Ph.D., 1967). In 1967 and 1968 he was assistant professor of biology at Idaho State University, then joined the faculty of the University of New Mex-

ico in 1968, becoming professor in 1977. Sandra Ligon is lecturer in biology at the University of New Mexico. She was graduated from Knox College in 1971 with a B.S. and obtained an M.S. from the University of New Mexico in 1973, then worked for two years for the U.S. National Fish and Wildlife Laboratory.

CHARLES G. SIBLEY and **JON E. AHLQUIST** ("Reconstructing Bird Phylogeny by Comparing DNA's"). Sibley earned a Ph.D. in zoology from the University of California at Berkeley in 1948, then held positions at San Jose State College and at Cornell University before joining Yale University in 1965 as a professor of biology. From 1970 to 1976 Sibley was director of the Peabody Museum of Natural History. In 1986 Sibley joined the faculty at San Francisco State University as the Dean's Professor of Science. Ahlquist has a B.S. in biology from Cornell and a M.S. and Ph.D. from Yale. From 1972 to 1977 he was a curatorial associate at the Peabody Museum. Ahlquist took his present position with the museum in 1977 and joined the faculty of Yale's biology department in 1981, where he is an associate research scientist and lecturer.

WILLIAM A. CALDER III ("The Kiwi") is professor of ecology and evolutionary biology at the University of Arizona. A graduate of the University of Georgia, he obtained his Ph.D. from Duke University in 1966. He taught at Duke and at the Virginia Polytechnic Institute before moving to Arizona in 1969. In addition to his university teaching and research tasks he is a life member and trustee of the Rocky Mountain Biological Laboratory. In 1977 he was visiting professor at the University of New South Wales in Australia.

PETER WELLNHOFER ("*Archaeopteryx*") is the main curator and vice-director of the Bavarian State Collection for Paleontology and Historical Geology in Munich. He received his Ph.D. in paleontology from the Ludwig Maximilians University in Munich in 1964. In 1972 he accepted an invitation from the Carnegie Museum of Natural History in Pittsburgh to conduct six months of research in the U.S. He is also the editor of the *Encyclopedia of Paleoherpetology*.

Bibliographies

1. Learning by Instinct

Curio, E., and W. Vieth. 1978. Cultural transmission of enemy recognition. *Science* 202:899–901.

Gould, J. L. 1982. *Ethology: The mechanisms and evolution of behavior*. W. W. Norton & Co.

Marler, P. 1984. Song learning: Innate species differences in the learning process. In *The biology of learning*, eds. P. Marler and H. S. Terrace. Springer-Verlag.

Marler, P., and H. S. Terrace, eds. 1984. *The biology of learning*. Springer-Verlag.

Gould, J. L. 1986. The biology of learning. *Annual Review of Psychology* 37:163–192.

Gould, J. L., and C. G. Gould. 1988. *The honey bee*. Scientific American, Inc.

2. Internal Rhythms in Bird Migration

Aschoff, J. 1980. Biological clocks in birds. In *Acta XVII Congressus Internationalis Ornithologici*, ed. R. Nohring. Verlag der Deutschen Ornithologischen Gesellschaft.

Gwinner, E. 1981. Circannual systems. In *Handbook of behavioral neurobiology*, vol. 4, ed. J. Aschoff. Plenum Press.

Mead, C. 1983. *Bird migration*. Facts on File Publications.

Gwinner, E. 1986. *Circannual rhythms*. Springer-Verlag.

Gwinner, E. 1986. Circannual rhythms in the control of avian migrations. *Advances in the Study of Behavior* 16:191–228.

Berthold, P. 1989. The control of migration in European warblers. In *Acta XIX Congressus Internationalis Ornithologici*, ed. H. Ouellet. University of Ottawa Press.

Gwinner, E. 1989. Photoperiod as a modifying and limiting fact in the expression of avian circannual rhythms. *Journal of Biological Rhythms* 4:237–250.

Gwinner, E., ed. 1990. *Bird migration: Physiology and ecophysiology*. Springer-Verlag.

Wiltschko, W., and R. Wiltschko. 1990. Magnetic orientation and celestial cues in migratory orientation. *Experientia* 46:342–352.

3. Memory in Food-hoarding Birds

Cowie, R. J., J. R. Krebs and D. F. Sherry. 1981. Food storing by marsh tits. *Animal Behaviour* 29:1252–1259.

Sherry, D. F., J. R. Krebs and R. J. Cowie. 1981. Memory for the location of stored food in marsh tits. *Animal Behaviour* 29:1260–1266.

Vander Wall, S. B. 1982. An experimental analysis of cache recovery in Clark's nutcracker. *Animal Behaviour* 30:84–94.

Shettleworth, S. J. 1984. Learning and behavioral ecology. In *Behavioural ecology*, 2nd ed., eds. J. R. Krebs and N. B. Davies. Blackwell.

Kamil, A. C., and R. P. Balda. 1985. Cache recovery and spatial memory in Clark's nutcrackers (*Nucifraga columbiana*). *Journal of Experimental Psychology: Animal Behavioral Processes* 11:95–111.

Shettleworth, S. J. 1985. Food storing by birds: Implications for comparative studies of memory. In *Memory systems of the brain*, eds. N. M. Weinberg, J. L. McGaugh and G. Lynch. Guilford.

Sherry, D. F., and D. L. Schacter. 1987. The evolution of multiple memory systems. *Psychological Review* 94:439–454.

Sherry, D. F., and A. L. Vaccarino. 1989. Hippocampus and memory for caches in black-capped chicadees. *Behavioral Neuroscience* 103:308–318.

Kamil, A. C., and R. P. Balda. 1990. Spatial memory in seed caching corvids. In *The psychology of learning and motivation*, ed. G. H. Bower. Academic Press.

Krebs, J. R. 1990. Food-storing birds: Adaptive specialization in brain and behaviour? *Philosophical Transactions of the Royal Society, Series B* 329:153–160.

Shettleworth, S. J. 1990. Spatial memory in food-storing birds. *Philosophical Transactions of the Royal Society, Series B* 329:143–151.

Shettleworth, S. J., J. R. Krebs, S. D. Healy and C. M. Thomas. 1990. Spatial memory of food-storing tits (*Parus ater* and *P. atricapillus*): Comparison of storing and non-storing tasks. *Journal of Comparative Psychology* 104:71–81.

Vander Wall, S. B. 1990. *Food hoarding in animals*. University of Chicago Press.

4. From Bird Song to Neurogenesis

Thorpe, W. H. 1958. The learning of song patterns by birds, with special reference to the song of the chaffinch (*Fringilla coelebs*). *Ibis* 100:535–570.

Nottebohm, F., and A. P. Arnold. 1976. Sexual dimorphism in vocal control areas of the songbird brain. *Science* 194:211–213.

DeVoogd, T. J., and F. Nottebohm. 1981. Gonadal hormones induce dendritic growth in the adult brain. *Science* 214:202–204.

Goldman, S. A., and F. Nottebohm. 1983. Neuronal production, migration and differentiation in a vocal control nucleus of the adult female canary brain. *Proceedings of the National Academy of Sciences* 80:2390–2394.

Paton, J. A., and F. Nottebohm. 1984. Neurons generated in the adult brain are recruited into functional circuits. *Science* 225:1046–1048.

Burd, G. D., and F. Nottebohm. 1985. Ultrastructural characterization of synaptic terminals formed on newly generated neurons in a song control nucleus of the adult canary forebrain. *Journal of Comparative Neurology* 240:143–152.

Nottebohm, F., M. E. Nottebohm and L. A. Crane. 1986. Developmental and seasonal changes in canary song and their relation to changes in the anatomy of song-control nuclei. *Behavioral and Neural Biology* 46:445–471.

Nottebohm, F., M. E. Nottebohm, L. A. Crane and J. C. Wingfield, 1987. Seasonal changes in gonadal hormone levels of adult male canaries and their relation to song. *Behavioral and Neural Biology* 47:197–211.

Alvarez-Buylla, A., and F. Nottebohm. 1988. Migration of young neurons in adult avian brain. *Nature* 335:353–354.

Alvarez-Buylla, A., M. Theelen and F. Nottebohm. 1988. Birth of projection neurons in the higher vocal center of the canary forebrain before, during and after song learning. *Proceedings of the National Academy of Sciences* 85:8722–8726.

Canady, R. A., G. D. Burd, T. D. DeVoogd, and F. Nottebohm. 1988. Effect of testosterone on input received by an identified neuron type of the canary song system: A Golgi/EM/Degeneration study. *Journal of Neuroscience* 8:3770–3784.

Nottebohm, F. 1988. Hormonal regulation of synapses and cell number in the adult canary brain and its relevance to theories of long-term memory storage. In *Neural control of reproductive function*, eds. J. M. Lakoski, J. Regino Perez-Polo and D. K. Rassin. Alan R. Liss, Inc.

5. The Hearing of the Barn Owl

Payne, R. S. 1971. Acoustic location of prey by barn owls (*Tyto alba*). *Journal of Experimental Biology* 54:535–573.

Konishi, M. 1973. How the owl tracks its prey. *American Scientist* 61:414–424.

Knudsen, E. I. 1980. Sound localization in birds. In *Comparative studies of hearing in vertebrates*, eds. A. N. Popper and R. R. Fay. Springer-Verlag.

Dooling, R. J. 1982. Auditory perception in birds. In *Acoustic communication in birds*, vol. I, eds. D. E. Kroodsma and E. H. Martin. Academic Press, Inc.

Knudsen, E. I. 1984. Synthesis of a neural map of auditory space in the owl. In *Dynamic aspects of neocortical function*, eds. G. M. Edelman, W. M. Cowan and W. E. Gall. John Wiley and Sons.

Knudsen, E. I., S. du Lac and S. D. Esterly. 1987. Computational maps in the brain. *Annual Review of Neuroscience* 10:41–65.

Knudsen, E. I. 1988. Sensitive and critical periods in the development of sound localization. In *From message to mind: Directions in developmental neurobiology*, eds. S. Easter, K. Barald and B. Carlson. Sinauer Associates, Inc.

Konishi, M., T. T. Takahashi, H. Wagner, W. E. Sullivan and C. E. Carr. 1988. Neurophysiological and anatomical substrates of sound localization in the owl. In *Auditory function*, eds. G. M. Edelman, W. E. Gall and W. M. Cowen. John Wiley and Sons.

6. Sexual Selection in Bowerbirds

Marshall, A. J. 1954. *Bower-birds: Their displays and breeding cycles*. Oxford University Press.

Gilliard, E. T. 1969. *Birds of paradise and bower birds*. Natural History Press.

Pruett-Jones, M. A., and S. G. Pruett-Jones. 1982. Spacing and distribution of bowers in Macgregor's bowerbird (*Amblyornia macgregoriae*). *Auk* 102:334–341.

Arnold, S. 1983. Sexual selection: The interface of theory and empiricism. In *Mate choice*, ed. P. Bateson. Cambridge University Press.

Borgia, G. 1985. Bowers as markers of male quality: Test of a hypothesis. *Animal Behaviour* 35:266–271.

———. 1985. Bower destruction and sexual competition in the satin bowerbird (*Ptilonorhynchus violaeceus*). *Behavioral Ecology and Sociolobiology* 18:91–100.

Borgia, G., S. G. Pruett-Jones and M. A. Pruett-Jones. 1985. Bowers as markers of male quality.

Zeitschrift für Tierpsychologie/Journal of Comparative Ethology 65:225–236.

Loffredo, C., and G. Borgia. 1985. Male courtship vocalizations as cues for mate choice in the satin bowerbird (*Ptilonorhynchus violaceus*). *Auk* 103:189–195.

Borgia, G., S. G. Pruett-Jones and M. A. Pruett-Jones. 1985. The evolution of bower-building and the assessment of male quality. *Zeitschrift für Tierpsychologie/Journal of Comparative Ethology* 67:225–236.

Borgia, G. 1986. Satin bowerbird parasites: A test of the bright male hypothesis. *Behavioral Ecology and Sociobiology* 19:355–358.

Borgia, G., and M. Gore. 1986. Sexual competition by feather stealing in the satin bowerbird (*Ptilonorhynchus violaceus*). *Animal Behaviour* 34:727–738.

Borgia, G., I. Kaatz and R. Condit. 1987. Flower choice and the decoration of the bower of the satin bowerbird (*Ptilonorhynchus violaceus*): A test of hypotheses for the evolution of display. *Animal Behaviour* 35:1129–1139.

Diamond, J. 1987. Bower building and decoration by the bowerbird *Amblyornis inornatus*. *Ethology* 74:177–204.

———. 1988. Experimental study of bower decoration by the bowerbird *Amblyornis inornatus*, using colored poker chips. *American Naturalist* 131:631–653.

Borgia, G., and K. Collis. 1989. Female choice for parasite-free male satin bowerbirds and the evolution of bright male plumage. *Behavioral Ecology and Sociobiology* 22:445–454.

7. The Birds of Paradise

Gilliard, E. T. 1969. *Birds of paradise and bower birds.* Natural History Press.

Beehler, B. 1983. Frugivory and polygamy in birds of paradise. *Auk* 100:1–12.

Beehler, B., and S. G. Pruett-Jones. 1983. Display dispersion and diet of birds of paradise: A comparison of nine species. *Behavioral Ecology and Sociobiology* 13:229–238.

Pratt, T. K., and E. W. Stiles. 1983. How long fruit-eating birds stay in the plants where they feed: Implications for seed dispersal. *American Naturalist* 122:797–805.

Diamond, J. 1986. Biology of birds of paradise and bowerbirds. *Annual Review of Ecology and Systematics* 17:17–27.

Beehler, B. 1987. Birds of paradise and mating system theory—predictions and observations. *Emu* 78:78–89.

Beehler, B., and M. S. Foster. 1988. Hotshots, hotspots, and female preference in the organization of lek mating system. *American Naturalist* 131:203–219.

8. Cooperative Breeding in the Acorn Woodpecker

MacRoberts, M. H., and B. R. MacRoberts. 1976. *Social organization and behavior of the acorn woodpecker in central coastal California.* American Ornithologists' Union. Monograph No. 21. Allen Press.

Brown, J. 1978. Avian communal breeding systems. *Annual Review of Ecology and Systematics* 9:123–155.

Emlen, S., and S. Vehrencamp. 1983. Cooperative breeding strategies among birds. In *Perspectives in ornithology: Essays presented for the centennial of the American Ornithologists' Union*, eds. A. Brush and G. A. Clark. Cambridge University Press.

Woolfenden, G. E., and J. W. Fitzpatrick. 1984. *The Florida scrub jay: Demography of a cooperative-breeding bird.* Princeton University Press.

Brown, J. L. 1987. *Helping and communal breeding in birds.* Princeton University Press.

Koenig, W. D., and R. L. Mumme. 1987. *Population ecology of the cooperatively breeding acorn woodpecker.* Princeton University Press.

Skutch, A. J. 1987. *Helpers at birds' nests.* University of Iowa Press.

Stacey, P. B., and J. D. Ligon. 1987. Territory quality and dispersal options in the acorn woodpecker, and a challenge to the habitat saturation model of cooperative breeding. *American Naturalist* 130:654–676.

Ligon, J. D., and P. B. Stacey. 1989. On the significance of helping behavior in birds. *Auk* 106:700–705.

Emlen, S. T. 1984. Cooperative breeding in birds and mammals. In *Behavioural ecology: An evolutionary approach*, 2nd ed., eds. J. R. Krebs and N. B. Davies. Sinauer Associates.

Stacey, P. B., and W. D. Koenig, eds. 1990. *Cooperative breeding in birds: Long-term studies of ecology and behavior.* Cambridge University Press.

Stacey, P. B., and J. D. Ligon. In press. The benefits of philopatry hypothesis for the evolution of cooperative breeding: Variation in territory quality and group size effects. *American Naturalist*

9. The Cooperative Breeding Behavior of the Green Woodhoopoe

Emlen, S. T. 1984. Cooperative breeding in birds and mammals. In *Behavioral ecology: An evolutionary approach*, 2nd ed., eds. J. R. Krebs and N. B. Davies. Sinauer Associates.

Brown, J. L. 1978. Avian communal breeding systems. *Annual Review of Ecology and Systematics* 9:123–155.

Ligon, J. D., and Ligon, S. H. 1978. The communal social system of the green woodhoopoe in Kenya. *Living Bird* 16:159–197.

Ligon, J. D. 1981. Demographic patterns and communal breeding in the green woodhoopoe, *Phoeniculus purureus*. In *Natural selection and social behavior: Recent research and new theory*, eds. R. D. Alexander and D. W. Tinkle. Chiron Press.

Koenig, W. D., and F. A. Pitelka. 1981. Ecological factor and kin selection in the evolution of cooperative breeding in birds. In *Natural selection and social behavior: Recent research and new theory*, eds. R. D. Alexander and D. W. Tinkle. Chiron Press.

Emlen, S. T. 1984. Cooperative breeding in birds and mammals. In *Behavioural ecology: An evolutionary approach*, eds. J. R. Krebs and N. B. Davies. Sinauer Associates.

Wiley, R. H., and K. N. Rabenold. 1984. The evolution of cooperative breeding by delayed reciprocity and queing for favorable social positions. *Evolution* 38:609–621.

Brown, J. L. 1987. *Helping and communal breeding in birds*. Princeton University Press.

Ligon, J. D., and S. H. Ligon. 1988. Territory quality: Key determinant of fitness in the group-living green woodhoopoe. In *The ecology of social behavior*, ed. C. Slobodchikoff. Academic Press.

Ligon, J. D., and S. H. Ligon. 1989. Green woodhoopoe. In *Lifetime reproduction in birds*, ed. I. Newton. Academic Press.

Stacey, P. B., and W. D. Koenig, eds. 1989. *Cooperative breeding in birds: Long-term studies of ecology and behavior*. Cambridge University Press.

Ligon, J. D. 1991. Cooperation and reciprocity in birds and mammals. In *Kin recognition*, ed. P. G. Hepper. Cambridge University Press.

10. Reconstructing Bird Phylogeny by Comparing DNA's

Sibley, C. G., and J. E. Ahlquist. 1983. The phylogeny and classification of birds based on the data of DNA-DNA hybridization. *Current Ornithology* 1:245–292.

————. 1985. The phylogeny and classification of the new world suboscine passerine birds (Passeriformes: Oligomyodi: Tyranni). In *Neotropical ornithology*, eds. P. A. Buckley, M. S. Foster, E. S. Morton, R. S. Ridgely and F. G. Buckley. American Ornithologists' Union Monograph No. 36.

Darnell, J., H. Lodish and D. Baltimore. 1986. *Molecular cell biology*. Scientific American, Inc.

Sibley, C. G., J. E. Ahlquist and B. L. Monroe, Jr. 1988. A classification of the living birds of the world based on DNA-DNA hybridization studies. *Auk* 105:409–423.

Stryer, L. 1988. *Biochemistry*, 3rd ed. W. H. Freeman and Company.

Sibley, C. G., and B. L. Monroe, Jr. 1990. *Distribution and taxonomy of the birds of the world*. Yale University Press.

Sibley, C. G., and J. E. Ahlquist. 1990. *Phylogeny and classification of birds: A study in molecular evolution*. Yale University Press.

11. The Kiwi

Cracraft, J. 1974. Phylogeny and evolution of the ratite birds. *Ibis* 116:494–521.

Rahn, H., and A. Ar. 1974. The avian egg: Incubation time and water loss. *Condor* 76:147–152.

Reid, B., and G. Williams. 1975. The kiwi. In *Biogeography and ecology in New Zealand*, ed. G. Kuschel. Dr. W. Junk B. V. Publishers.

Calder, W. A., III. 1979. The kiwi and egg design: Evolution as a package deal. *Bioscience* 29:461–467.

Rahn, H., A. Ar and C. V. Paganelli. 1979. How bird eggs breathe. *Scientific American* 240:46–55.

DeBoer, L. E. M. 1980. Do the chromosomes of the kiwi provide evidence for a monophyletic origin of the ratites? *Nature* 287:84–85.

Silyn-Roberts, H. 1982. The pore geometry and structure of the eggshell of the North Island brown kiwi, *Apteryx australis mantelli*. *Journal of Microscopy* 130:23–36.

Calder, W. A., III. 1984. *Size, function and life history: Adaptive strategies*. Harvard University Press.

Gould, S. J. 1986. Of kiwi eggs and the liberty bell. *Natural History* 11:20–29.

Sotherland, P. R., and H. Rahn. 1987. On the composition of bird eggs. *Condor* 89:48–65.

Vleck, C. M., and D. Vleck. 1987. Metabolism and energetics of avian embryos. *Journal of Experimental Zoology*, Supplement 1:111–125.

McLennan, J. A. 1988. Breeding of the North Island brown kiwi, *Apteryx australis mantelli*, in Hawke's Bay, New Zealand. *New Zealand Journal of Ecology* 11:89–97.

Calder, W. A., III. 1990. The kiwi and its egg. In *The kiwi*, eds. E. Fuller and R. Harrish-Ching. Seto Publishing.

Rowe, B. 1978. Incubation temperatures of the North Island brown kiwi. *Notornis*

12. *Archaeopteryx*

Heilmann, G. 1926. *The origin of birds*. Appleton.

Parkes, K. C. 1966. Speculations on the origin of feathers. *Living Bird* 5:77–86.

Regal, P. J. 1975. The evolutionary origin of feathers. *Quarterly Review of Biology* 50:35–66.

Ostrom, J. H. 1976. *Archaeopteryx* and the origin of birds. *Biological Journal of the Linnean Society* 8:91–182.

Feduccia, A. 1980. *The age of birds*. Harvard University Press.

———. 1984. *It started in the Jurassic: The fascinating phylogeny of birds*. Gerstenberg Verlag.

Peters, D. S. 1984. Konstruktionsmorphologische Gesichtspunkte zur Entstehung der Vögel. *Natur und Museum* 114:119–210.

Hecht, M. K., J. H. Ostrom, G. Viohl and P. Wellnhofer, eds. 1985. The beginnings of birds. *Proceedings of the International Archaeopterx Conference, Eichstätt, 1984*. Freunde des Juramuseums.

Cracraft, J. 1986. The origin and early diversification of birds. *Palaeobiology* 12:383–399.

Padian, K., ed. 1986. The origin of birds and the evolution of flight. *Memoirs of the California Academy of Sciences* 8:1–98.

Wellnhofer, P. 1988. A new specimen of *Archaeopteryx*. *Archaeopteryx* 6:1–30.

Sources of the Photographs

Stephen Dalton, Photo Researchers, Inc.: Figure 1.1

Stephen B. Vander Wall, Utah State University: Figure 3.11

Fernando Nottebohm: Figure 4.4

Eric I. Knudsen, Stanford University School of Medicine: Figure 5.1

Hans and Judy Beste, Animals, Animals: Figures 6.2 and 6.4

Bruce M. Beehler: Figures 7.1, 7.4 (*right*) and 7.6

David Gillison, Peter Arnold, Inc.: Figure 7.4 (*left*)

G. Ziesler, Peter Arnold, Inc.: Figure 7.5

Walter D. Koenig, University of California at Berkeley: Figure 8.2

Cynthia Carey, University of Colorado: Figure 11.10

Peter Wellnhofer: Figures 12.1 and 12.3

Renate Liebreich, Bavarian State Collection for Paleontology and Historical Geology: Figure 12.4

INDEX

Page numbers in *italics* indicate illustrations.